Felix Leopold Oswald

Zoological Sketches

A contribution to the out-door study of natural history

Felix Leopold Oswald

Zoological Sketches

A contribution to the out-door study of natural history

ISBN/EAN: 9783337095246

Printed in Europe, USA, Canada, Australia, Japan

Cover: Foto ©berggeist007 / pixelio.de

More available books at **www.hansebooks.com**

ZOOLOGICAL SKETCHES

A CONTRIBUTION

TO THE

OUT-DOOR STUDY OF NATURAL HISTORY.

BY

FELIX L. OSWALD,
AUTHOR OF
" SUMMERLAND SKETCHES OF MEXICO AND CENTRAL AMERICA."

" The Book of Nature is ever new, though never self-conflicting."—LESSING.

WITH THIRTY-SIX ILLUSTRATIONS
BY HERMANN FABER.

LONDON:
W. H. ALLEN & CO. 13 WATERLOO PLACE.
1883.

PREFACE.

The tendencies of our realistic civilization make it evident that the study of natural science is destined to supersede the mystic scholasticism of the Middle Ages, and I believe that the standards of entertaining literature will undergo a corresponding change. The Spirit of Naturalism has awakened from its long slumber.

A year after the birth of the Emperor Tiberius, says Plutarch, a Grecian trading-vessel sailed along the coast of Ætolia in the Gulf of Patras, and when the sun went down the crew assembled at the helm to while away the night with songs and stories. The night was calm, and some of the sailors had already fallen asleep, when they heard from the coast a loud voice calling the name of their steersman, Thamus. They were all struck dumb with amazement, but at the third call Thamus manned himself and answered with a loud mariner's shout.

"O Thamus," the voice called again, "when you reach the heights of Palodes announce that the great Pan is dead!"

Four hours later, when the moonlit hills of Palodes

hove in sight, Thamus complied with the strange request, and a minute after, the coast resounded with indescribable shrieks and lamentations that continued for a long time, till they finally died away in the heights of the Acarnanian Mountains.

The tradition bears the mark of that suggestiveness which distinguishes a philosophical allegory from a priest-legend. Pan was the God of Nature. Can Plutarch have divined the significance of the impending change? Whatever is natural is wrong, was the keystone dogma of the mediæval schoolmen. The naturalism of antiquity was crushed by supernatural and antinatural dogmas. The worship of joy yielded to a worship of sorrow, the study of living nature to the study of dead languages and barren sophisms. Literature became a farrago of ghost-stories, monks' legends, witchcraft- and miracle-traditions, and astrological vagaries. The poison of antinaturalism tainted every science and every art and perverted the very instincts of the human mind. Painters vied in the representation of revolting tortures. The exiles of Mount Parnassus assembled on Mount Golgotha. The moralists that had suppressed the Olympic festivals compensated the public with autos-da-fé. The whole history of the Middle Ages is, indeed, the history of a long war against nature.

But nature has at last prevailed. Delusions are clouds, and the storm of the Thirty Years' War has cleared our sky. The real secret of the astounding success of modern science and industry is a general *renaissance* of naturalism, and the same revival begins to manifest its influence in the tendencies of modern literature. Ghost-stories are going out of fashion. Like scrofula and other bequests of the Middle Ages, the sickly pessimism of the sentimental school is yielding to the influence of a revived taste for the pleasures of out-door life. Books of travel, of sports and adventure, historical, zoological, and even biological and cosmological studies, are fast superseding the historical romances of the last generation. Even the Pariahs of our reading-rooms have advanced from ghost-hunts to scalp-hunts, from impossibilities to improbabilities. And, moreover, the progress of natural science tends to supersede fiction by making it superfluous—even for romantic purposes. There is more romance in the travels of Humboldt, more magic in the idyls of Thoreau and the revelations of Darwin and Haeckel, than in all the fancies of the mediæval miracle-mongers. The wonders of nature begin to eclipse the wonders of supernaturalism. A Zoological Garden attracts more sight-seers than the best Passion-play. Pan has revived.

The plan of the present volume is modest enough: its theories are mere suggestions; its limits have often obliged me to reduce a chapter of zoological adventures to a page of zoological anecdotes. But in offering it as a contribution to the entertaining literature of the English language, my diffidence arises from a distrust in my own abilities rather than from the deficient interest of the subject itself, for the history of that literature has repeatedly proved that natural science can be made more attractive than the products of fiction or mysticism—by just as much as the resources of nature exceed the resources of her rivals.

FELIX L. OSWALD.

CINCINNATI, *March*, 1882.

CONTENTS.

CHAPTER I.

OUR FOUR-HANDED RELATIVES.

PAGE

Human Affinities—A Biological Problem—Nose-Apes—The Forehead-Criterion—Delusive Symptoms—The Wanderoo—A Man and Brother—The Medicean Paragon—Curious Analogies—Scratching and Striking Arguments—Monkey-Habits—The Marvels of the Brain—Singular Contrasts—Untamable Brutes—Amativeness—The Spider-Monkey—African Baboons—A Bold Marauder—Superhuman Fists—A Vegetarian Argument—Capuchin-Monkeys—The Ne Plus Ultra of Cowardice—Obstreperous Passengers—Frederick Gerstaecker's Expedient—Dwarf Monkeys—Midas Rosalia—The Moor-Ape—Total Depravity—A Farmers' Pest—Monos de Cadena—The Genius of Mischief—A Fatal Oversight—Platonic Homunculi—The Pet of Cartagena—Reasoning Capacities—Dogs and Monkeys—A Four-handed Buddhist—Nest-building Animals—Strange Bedfellows—A Picnic Adventure—Lesser Evils—A Practical Physiognomist—Defying the Landlord—The Macacus Radiatus—Self-Reliance—The Scale of Intelligence—Curiosity—The Secret of Epaminondas—An International Language—The Red Howler—Treetop Serenades—The Science of Tucbeer—Anthropoid Apes—Pansy's Stratagem—Love at First Sight—The Advantages of Circumspection—The Javanese Manki—Mind *vs.* Matter—A Casus Belli—Victorious Impudence—Yielding under Protest—Esprit de Corps—A Self-Asserting Pet—Catching a Tartar—Captain Hess—Salto Mortale—Freebooters—The Patron Saint of

Thieves—Incurable Kleptomaniacs—Mr. Thielman's Cook—Misplaced Confidence—Monkey Education—A School for Pickpockets—Unnatural Mothers—Living by Stealth—Nest-hiding—A Queer Predilection—Buddha's Foibles—A Moral Experiment—Martyrs to Free Inquiry—Taciturn Monkeys—The Rhesus Baboon—An Obscene Saint—Four-handed Drunkards—The Force of a Bad Example—Is our Love of Salt Natural?—Original Sin—Eating on Principle—Fruges Consumere Nati—Lung-Poison—A Curious Experiment—What Kills our Menagerie Monkeys—A Physiological Puzzle—Anthropoid Monkeys—Their Antipathy to Children—Paradoxical Character-Traits—The Monkeys of New Freiburg—Over-practical Jokes—Fatal Consequences—The Victim of the Wanderoos—Vicarious Atonement—The Faculty of Dissimulation. 21

CHAPTER II.

MOUNTAIN SHEEP.

By-Laws of Nature—Man and his Fellow-Creatures—Solution of a Zoological Mystery—Survival under Difficulties—The Haunts of the Mountain Sheep—Grazing Cimarróns—Picket-Posts—Retreat-Tactics—Adventure of an American Engineer—A Camp of Observation—The Progress of Culture—Four-footed Emigrants—Colonel Pennypacker's Recollections—Our Last Hunting-Grounds—The Guests of the Northland—Winter-Horrors—A Strange Honeymoon—Elder Millard's Discovery—Protective Instincts—An Ill-fated Rookery—The Gamekeeper of Rheinharts-Brunn—Hibernation—Biornir-Nott — Weather-Prophets — A Storm-Signal — Highland Camps—Zoological Nondescripts—Perplexities of a Modern Naturalist—Untamable Kids—A Dangerous Pet—Don Panchito—Intemperate Quadrupeds—Borracheria—The Perils of Pulque—A Long-horned Dilemma—Ex Infernis—A Singular Instinct—Family Duties—Væ Victis—A Last Resort—Zoological Superstitions—How Wild Sheep descend a Precipice—Sheep-Hounds—A Hunter's Ruse — Domestic *Feræ* —" Sheep-tickle" —The Wind River Range—Lonely Hunting-Grounds—The Meadows of the Sierra de San Simon—Four-footed Bachelors—A Wary Hermit—Rock-Labyrinths—An Ethnological Conjecture 60

CHAPTER III.

A STEP-CHILD OF NATURE.

PAGE

Mexican Mountain-Forests—The Mono Espectro—Voices of the Wilderness—Dog-day Siesta—A Daylight Owl—The Tardo Negro—A Chase in the Tree-tops—Constitutional Stoicism—The Better Part of Valor—An Awkward Predicament—Frugal Habits —A Child of the Air—Arboreal Mammals—Euphorbia-Trees—A Hairy Hamadryad—Four-footed Trappists—A Vital Problem— The Crime of Helplessness—Adventure of a Lumberman—Martes torquatus—Chicken-Thieves—Mischief-Joy—Life under Difficulties—The Tarda Morena—In Articulo Mortis—A Sleepless Creature—Survivors of the Fern-Age—The Spotted Sloth—A Three-legged Pensioner—Grunts et præterea nihil—A Freak of Nature —Imperturbable Stoicism—A Philosopher of the Horatian School —Cinderella—A Strange Pet—Abusing Good-Nature—Precocious Egotism—Nostalgia—Universal Instincts — Non-resistance—The Wisdom of Compensating Nature—Secrets of Happiness . . 79

CHAPTER IV.

SECRETIVENESS.

Rudimentary Instincts—Their Natural Development—The Faculty of Direction—Hunting under Water—The Simia Destructor and his Four-footed Rivals—Phrenological Indications—The Skull of a Weasel—The Hiding-Faculty—Artful Dodgers—Nest-Hiding— Bears and Birds—The Orchard Oriole—Birds Fooling their Pursuers—Instincts of Young Birds—Impromptu Hiding-Place—An Astonished Fox-Hunter—Procul de Jove—Outwitting his Landlord —Raccoons and Muskrats—What becomes of Dead Animals— Secretiveness in Articulo Mortis—A Lost Pet—Where they found Him—Night-Walkers—The Ghost Hour—Mysteries of a Poultry-House—Night Visions—Migratory Birds—Their Favorite Routes— Nocturnal Wanderers—Paso del Norte—Plenty Room higher up

PAGE

—Chronological Instinct—Sunday in France—Sanguinary Sabbaths—The Virginia Partridge—Cautious Marauders—Rat-Patriarchs—Ratification-Meeting—Wall-Mice—Discovery of a Boarding-house-Keeper—Secret Lodgers—The American Skunk—Hidden Headquarters—Extinct Animals—Our Natural Game—Preserves—Panthers and Wolves—A Beast Asylum—The Mountains of North Carolina—French Wolves—The Mystery of Allendorf—A Strange Spoor—In the Salpetar-Loch—The Oberförster's Opinion—Science vs. Empiricism—A Zoological Lecture—Stubborn Sceptics—An argumentum ad hominem—"Secret Camelopards"—A Frank Forester 97

CHAPTER V.

BATS.

Curious Fossils—A Relic of a Bygone World—Children of Tartarus—The Winged Lemur—Night-Apes—The Bat-Mystery—Spallanzani's Conjecture—The Sixth Sense—Night-Walkers—Gluttons and their Characteristics—Bat-Voices—Aristotle's Opinion—A Winged Nurse—Useful Hooks—Winter-Quarters—The Effects of Frost—Latent Vitality—Queer Dormitories—The Grottos of Posilippo—The Biels-Höhle—A Mass-Meeting House—Canadian Bats—Le Borgne Corné—Phrenological Reflections—Brainless Brutes—A Subterfuge—The Salzburg Acropolis—A "Bat-Rookery"—Routing his Lodgers—An Acherontic Spectacle—Children of Chaos—National Superstitions—Natt-Backa—Devil-Birds—A Paragon of Ugliness—Slandered Benefactors—Bacon-fat—The Cockatrice—Vampire-Lore—Ominous Symptoms—Sleeping under Difficulties—The Ghoul-Bat—A Vampire Trap—Bonpland's Receipt—A Valuable Accomplishment—Entomological Enigmas—Mosquitoes—How they subsist in the Woods—The "Sunken Lands"—An Ugly Dilemma—The Vampirus Spectrum—A Miraculous Instinct—Two Drunken Sailors—Exsanguis—Tropical Bats—The Kalong Nuisance—Voracity of the Javanese Roussette—Insatiable Boarders—A Tough Constitution—The Pets of Cape Angol—Rydenberg—Tropical Sports—The Dutch Colonists—A Fox-Chase in the Air—The Natives of Wynkoop's Bay—Fighting the Harpies—The "Sky-Fox"—Monkey-Birds—The Wages of Sin . . . 114

CHAPTER VI.

SACRED BABOONS.

PAGE

Waterton's Experiment—An Asylum for Birds and Beasts—The Children of the All-Father—Hindoo Ethics—Queer Protégés—Four-handed Demi-Gods—The Bhunder Baboon—Sacred Crocodiles—Eupeptic Pets—Sir Emerson Tennent's Statistics—Inviolate Carnivora—Deva-Ghee — Monkey-Hospital—High-Caste Apes—Strange Bequests—Pariah Monkeys—The Favorites of Brahm—Lunch-Fiends—Dangerous Superstitions—Captain Elphinstone's Gardener—The Lex Talionis—Sacred Bulls—Their Expensive Privileges—Reverend Quadrupeds—Fraternity and Equality—Dr. Vanjorden's Experience—Baboon Asylums—Punctual Boarders—The Pets of the Dhevadar—Cheek—A Natural Knapsack—The Papio Rhesus—Obstreperous Pensioners—A Council of War—Willy-nilly—The Mahakund—" Pious and Continent Paupers"—Race-Prejudices—A Hindoo Legend—Ravan and the Rishis—The Monkey-Honuman—Honuman's Stratagem—Unforeseen Results—The Sacred Mountain Lake—Incontestable Proofs—A Valuable Relic—The Spoils of the Virey—Inviolate Guests—The Saints of Khunar—Precautions—Impious Britishers—" Ludere cum Sacris"—Tempting the Saints—A Repentant Sinner—Plethoric Pensioners—Exiled Aristocrats—The Victims of the Sepoy Insurrection — Destitute Monkeys — City-Monkeys — A Four-handed Tramp—The Upper Ten—No False Modesty—Eccentric Mussulmans—Their Hatred of Idols—Resolute Mendicants—Kleptomania —The Delhi Bhunders—Race-Instincts—A Boy protected by Monkeys—The Cause of the Indian Insurrection—Interceding for a Pickpocket—Dr. Vanjorden's Servant—Mons. Duvancel's Mistake—Stuffing a Saint—" Wicked Harbarat's Place"—A Parthian Shot—Mohammedan Allies—Shah Allum's Sentence—Four-handed Convicts—Cemetery Monkeys— Dr. Mackenzie's Scrape — The Limits of Human Patience—A Question of Casuistry—Lecturing a Bull—Gymnastic Exploits 132

CHAPTER VII.

ANIMAL RENEGADES.

PAGE

Nature *vs.* Slavery—Secret Protestantism—Our Truant Pets—Dogs—Their Night-Rambles—Private Business—Proofs Positive—A Practical Argument—Gadding Cats—How Tomcats spend their Summer Vacations—Bush-Pork—The Goats of the Tyrolese Alps—Animal Renegades—Rebellion *en masse*—Wild Horses and Cows—Ownerless Dogs—Cabras Pardas—Wild-Cats and Feld-Cats—Bactrian Camels—The Wild Asses of Yemen—Burkhardt's Conjecture—Reappearance of old Race-Habits—The Khelp el Khamr—Eating a Sheik—My Mexican Friend—El Perro pelon—Tramp-Dogs—Business Practice—A Provident Puppy—Treasure-trove—Buddhistic Ethics—Broomstick Logic—A Declaration of Independence—Pampa Curs—The Canis Azaræ—Sierra Goats—Spontaneous Reversion—The Wild Cattle of the Brazos—Le Bidet Sauvage—An Equine Outlaw—Adventure in the Sambre Highlands—The Price of Liberty—Nemesis—Syrian Dogs—Dr. Tanner's Rivals—Chances for a Sausage-maker—Black-muzzled Cows—Hyena-Heads—A Singular Character-Trait—Mobbed by Mustangs—Wild Camels—Dangerous Travelling—Esprit de Corps—The Chinaco's Dog—Revenge by Proxy—A Werewolf's Den—Facing his Foes—An Unequal Combat—La Mort sans Phrase—Carrion-Eaters—An Unnatural Appetite—Dietetic Experiments—Communistic Insurrections—The Autocrat of the Animal Kingdom—Captain Kellerman—Misplaced Confidence—Into the Jaws of Death—An Appeal for Charity—Obstreperous Beggars—A Fatal Mistake—Orphan Puppies—The Tramp-Bitch—Foster-Children—An Errand of Mercy—Nocturnal Visits—Higher Duties . . 159

CHAPTER VIII.

PETS.

Natural Selection—Non-egotistical Instincts—Duties in Disguise—The Purpose of the Pet-Mania—Animals in Danger of Extermination—Natural Safeguards—Du Chaillu's Gorilla—Contradictory Reports—The Protégé of the Berlin Aquarium—Young Animals —Their Natural Tameness—Jaguar Cubs—Unconditional Surrender—Dietetic Influences—Infidel Pets—The Postmaster of San Pablo—Juanita—A Friend in Need—Jacko's Adventure—By the Skin of his Teeth—Independent Youngsters—Monkey-Babies— Their Ridiculous Tameness—My Bonnet-Macaque—A Four-handed Micawber—Love in Abstracto—Billy Hammock's Pets—A Fawnfinder—Dutch Storks—Their Fondness for Human Society—Foregoing their Winter-Trip—Rival Pets—Scared Chickens—Untamable Animals—Hospitality—Herr Haman's Boarders—The Guests of Miss Meiringer—Suaviter in modo—The Funeral of St. Renaldus—Grateful Bucks—The Santon of the Bakony Wald—Personal Magnetism—The White Doe of Rylstone—A Four-legged Kidnapper— An Affectionate Lynx — The Love-lorn Dolphin — A Strange Legend—The Whelps of the She-Wolf—A Story from India—The Wolf-Boys—Discovery of a Tax-Collector—Dietetic Predilections—A Strange Orphan—The Mountain-Wolf—Hunting Panthers and Trained Eagles—The Art of Falconry—Winged Retrievers—The Eagle of Judenburg—A Useful Bird—The Pensioners of Vishnu—St. Anthony's Pigs—An Eccentric Lady—The Curiosities of Mount Morris—Domestic Bears—Frank Buckland's Rats—Useless Pets are the most Affectionate—The Instinct of Freedom—Its Unexpected Revivals—The German Barnum—A Lion at large—Homeward Bound—Passive Resistance—Taming a Jackal—A Heroic Cure—The Canopy of a Beast-Tamer—A Reluctant Boarder—The Landlord of Eluelen—Love's Labor Lost— Snake-Charmers—Dr. Grotius's Remark—East Indian Beast-Charmers—Dangerous Pets—The Guruwalla—Lord Dalhousie's Wizard — The Coluber Dryas — Natural Magic—An Alligator-Charmer—Business Secrets—Professional Rat-catchers—East Indian Exiles—Gypsy Tricks—The Story of the Pied Piper—Orpheus —Self-sacrificing Animals—Major Keogh's Old Roan—The Train of the Wahabees—Trusty Dogs—The Shepherds of the Transvaal— Professor Schomberg's Experiment—The Power of Conscience— Enfant Perdu—The Lex Talionis—Micheline—Freak of a Tame Elephant—The Influence of Education—Hunting Panthers—A Test of Loyalty—Town-Dogs—Their Trials and Temptations— The Salzburg Acropolis—Sensitive Bats—Schopenhauer's Theory

—Musical Brutes—After Dark—Night-Terrors—Types and Archetypes—Origin of a Strange Character-Trait—The Power of Habit—Mill-Horses—Survival of the Fittest—The Influence of Domestication—A Conjecture 184

CHAPTER IX.

TRAPS.

Tool-making Animals—A First Attempt—Monkey-Men—The Value of Experience—Experts and Amateurs—Von Tschudi's Anecdote—A Vicuña-Trap—Hunter's Secret—Race-Foibles—How Minks are Trapped—Turkey-Pens—A Strange Fact—Secret Burrows—Favorite Haunts—Ideal Luxuries—Cardinal Retz—Partridge-Hunters—A Fatal Foible—Simple Rat-Traps—The Nest-hiding Faculty—Persistent Intruders—A Rat-proof Building—Otter Burrows—Fun-loving Animals—Unexpected Results—A Trapper's Trick—Olfactory Predilections—Strange Preferences—De Gustibus, etc.—Sewer Studies—Rat-Poison—A Singular Fact—The Language of Signs—Profiting by Experience—Rats—Their Parental Solicitude—The Trappers of Singapore—Outwitted Monkeys—Fuddle-Cakes—Dangerous Munificence—A Counter-Stratagem—In Gloria—The Wages of Sin—Steel-Traps—Ferocious Captives—Catching a Wild-Cat—Liberty or Death—Pitfalls—A Daring Trapper—In for it—American Hindoos—The Monkeys of Michoacan—A Test of Patience—Retribution by Proxy—Capuchin-Traps—The Trampa—Strong Inducements—The Inhumanity of Man to Man—Tempted Guests—A Monkey Patriarch—Impulse vs. Principle—A Luckless Stranger—Panic-Stricken—Lord Bacon's Remark—The Owl-Trap—Biters Bit—A Curious Experiment 223

CHAPTER X.

FOUR-FOOTED PRIZE-FIGHTERS.

Circus Combats—Herbert Spencer's Remark—Saxons and Latins—An Easy Way to cross the Styx—Death in Battle—Torres Vedras

CONTENTS.

—A Premature Boast—Rowdy Jack—The Scythe Brigade—Natural Selection—Ethical Contradictions—Bishop Riley's Respondent—An Argumentum ad Judicium—Gran Matanza—The Roman Circus Games—Their Magnitude and Influence—Startling Statistics—An Army of Wild Beasts—African Novelties—The Spanish Moriscos—Bull-Fights—A National Mania—Papal Edicts—Bulls and Counter-Bulls—José Perez—His Popularity and Successful Career—Popular Bulls—A Reluctant Triumphator—Apis-Worship—Fighting Elephants—No Sinecure—The Hutti—Untimely Tantrums—Dying Revenge—Hindostan Beast-Fights—Royal Sportsmen—The Prince of Baroda—A Private Menagerie—Famous Fighters—A Carnivorous Horse—Black Jan—The Idol of Samarang—Wolf-Baiting—Hungarian Sports—Sunday Morning Amusements—Bruin and the Pinchers—A Magnanimous Victor—Grizzly Bears—Mr. Presswood's Boy—The Mustela Martes—Guerre à l'outrance—The Rock-Plover—Paid in his own Coin—Ferrets and Rats—Bulldog Courage—Baron Gaisner's Wager—Tackling a Panther—An Astonished Dog—Prehistoric Sports—Spanish Man-hunts—Leonicico—A Werewolf—Balboa's "Adjutant"—A Rival of Cerberus—Aragon Hounds—Væ Victis—The Rhamadan—A Strategic Suggestion—Dutch Sports—The Pets of Amsterdam—Muidenhaven—Private Circenses—Dog-Dynasties—An Invincible Quadruped—Street-Fights—Not Easy to Scare—A Stranger from Ceylon—King Klaas 237

LIST OF ILLUSTRATIONS.

		PAGE
1.	Cimarrón Dogs	*Frontispiece.*
2.	The Chacma Baboon	25
3.	Total Depravity	29
4.	Unrequited Love	39
5.	Salto Mortale	46
6.	Misplaced Confidence	48
7.	Martyrs to Free Inquiry	52
8.	Winter Quarters	64
9.	A Steep Alternative	71
10.	A New Departure	85
11.	Prepaying the Debt of Nature	87
12.	A Slothful Family	89
13.	A Vantage-Ground	101
14.	Reconnoitring	105
15.	Sunday Morning	107
16.	Children of Erebus	117
17.	A Vampire-Trap	123
18.	A Fox-Chase in the Air	127
19.	The Pets of the Mahakhund	137
20.	Four-handed Lazzaroni	147
21.	The Limits of Human Patience	155
22.	Bactrian Camels	163
23.	Mustang Cows	171
24.	Wild Dogs	175
25.	"Juanita"	187
26.	Strange Messmates	193
27.	The Ger-Eagle	197

28. The Alligator-Charmer	.	207
29. A Dangerous Playmate	.	215
30. Survival of the Fittest	.	219
31. The Wages of Sin	.	229
32. "In for it"	.	233
33. Decoy Owls	.	235
34. A Reluctant Triumphator	.	243
35. The Rajah's Pet	.	247
36. "Væ Victis"	.	259

ZOOLOGICAL SKETCHES.

CHAPTER I.

OUR FOUR-HANDED RELATIVES.

OUR nearest relatives in the large family of the animal kingdom are undoubtedly the frugivorous four-handers, with some of their nocturnal congeners, but it would be difficult to classify the Quadrumana after the degree of that relationship: no naturalist could name the most man-like ape. It is a *reticulated*, rather than a graduated system of affinity, as Carl Vogt expresses it: the type of the human form is a centre from which the connecting lines diverge in various directions. To every supposed characteristic of our physical structure some genus or other of the multiform family has been found to exhibit a parallel; only the combination of these attributes distinguishes man from all monkeys.

The Latin word *simia* is derived from *simus* (flat-nosed), and Ælian considered the prominence of the human nose as a prerogative of our species; but Sir

Stamford Raffles discovered a nose-ape, the Bornean representative of the genus *Semnopithecus*, a big, long-tailed brute with a truly Roman proboscis and the narrow nostrils of the Caucasian race. In proportion to his size the white-handed capuchin-monkey of Western Guiana has a higher forehead than the two-legged inhabitants of his native woods; and the anatomist Camper demonstrated that with respect to the length of the tail-bones immortal man forms the connecting link between the lower apes and the orangs. The Arabs who question the human pedigree of the beardless Ethiopian would have to hail the wonderoo as a man and brother; and the male orang-outang, too, can boast of a chin-tuft that would do credit to a modern senator. With the exception of her expressive eyes, the face of the female orang is the most outrageous caricature of the Medicean paragon; man-like lineaments are, indeed, by no means a characteristic of the higher apes, and in that respect, at least, some of the macaques and Colobi would perhaps be the true anthropoids; but even the grotesque physiognomies of the South-American flat-noses are always redeemed by some strikingly human feature. The skinny spider-monkey has a mignon mouth and delicate white teeth, the little marmoset parts its hair in the middle, and the red howler (*Mycetes ursinus*) has the ear of a Spanish *maya*,—every fold, every dimple of the rim, a perfect fac-simile of the corresponding parts of the human auricle.

A similar analogy surprises the observer of certain gestures and tricks that distinguish our four-handed cousins from all lower animals. It is an innate habit of the Siamese gibbon to screen his eyes with the palm of his hand when looking at some distant object. Children in such an attitude often lean forward, and so does the gibbon,—as if a difference of three or four inches would avail him at a distance of a mile. Monkeys never grin without a twinkling movement of their eyelids. That might be caused by an interaction of the facial muscles; but what makes them avert their eyes if they pout, and stretch out their open hands if they surrender at discretion? Or why does the Rhesus monkey clutch his ears when he expects a hard blow? Does instinct teach him what his science has taught the anatomist,—viz., that the zygomatic arch is the weakest part of the skull? Or is it a result of educational influences, since the female of the same species is very apt to enforce her maternal authority by striking arguments? Peculiarities of structure may partly account for the singular tricks of certain species of monkeys. One of my acquaintances has caged a spaniel with a little long-fingered macaque, and at meal-times the monkey often resorts to a favorite stratagem of small boys in their scuffles with a bigger playmate, by looking sideways and keeping his hand rather out of sight when he is going to make a sudden grab. The dog knows that trick, and is all suspicion; but his twenty sharp teeth cannot compete with the twenty fingers of his little rival. Next to the eye, the prehensile hand is, indeed, the organic masterpiece of the Creator.

But stranger than the most fearfully-wonderful organism is the human mind, that mysterious medley of conflicting propensities, as Schopenhauer calls it; and the mental characteristics of our Darwinian relatives exhibit a not less wonderful diversity. The sundry breeds of our domestic dog differ considerably in talents and disposition; but that difference almost disappears before the character-contrasts of the various four-handers. The above-mentioned red howler of the Orinoco Valley is all but untamable, a most spiteful, morose, and repulsive brute; but his countryman the coaita or black spider-monkey is more absurdly affectionate than the fondest lap-dog. Solitary confinement almost breaks his heart; restored to liberty, he lavishes his embraces alike on friend and foe, and, *faute de mieux*, will hug an old tom-cat for hours together. Spurzheim's nomenclature has no word for that peculiar propensity; it has nothing to do with amativeness, nor is it "friendship," for it can dispense with reciprocation: it is rather an excess of affectionate confidence in the abstract, combined with a total want of resentment, for fear itself will not prevent the coaita from pressing his endearments upon an ill-tempered keeper.

A very different kind of confidence is that of the chacma baboon, who enters the fields of the Namaqua

Hottentots in broad daylight and often before the eyes of their fraus and children. After stuffing his cheek-

THE CHACMA BABOON.

pouches, he retreats, but leisurely and slowly, well knowing that no dog will dare to encounter him in the open fields. His eye-teeth are three inches long, and as sharp as those of a panther, but he rarely makes use of them; he relies on his arms, on the grasping and wrenching power of his superhuman fists. The orang-outang resembles him in this respect, but the orang never fights as long as he can possibly escape; the chacma yields to nothing but fire-arms, and finishes his meal in the presence of a troop of yelling children and yelping curs with the leonine calmness of a mastiff among a swarm of Skye terriers.

His intrepidity seems to refute the favorite argument of the anti-vegetarians, for the chacma subsists on berries, roots, and field-fruits; but with the same diet, and, in regard to its climate at least, a very similar habitat, the white-faced capuchin (*Cebus leucomeros*) is relatively and absolutely the greatest coward in creation: the mere sight of an unknown object is enough to frighten him into a fit of extravagant jumps and contortions. Cowardice is hardly the right word: if his conduct in captivity can be accepted as a criterion of his mental constitution, the Cebus seems to pass his life in a delirium of abject terror with rare and short self-possessed intervals. The screams that accompany his fits of trepidation make him a rather undesirable pet, for the constant exercise of his vocal apparatus has developed that organ to a degree out of all proportion to the size of the little alarmist. Frederick Gerstaecker, who shipped a boxful of these creatures on a Hamburg steamer, had to spend all his loose cash in *trinkgeld* to save his *protégés* from being kicked overboard by the exasperated crew. But, according to Montaigne, poltroonery is merely a sign of unusual foresight; and, if this be true, the providential faculties of the capuchin must amount almost to clairvoyance.

In his lucid moods the Cebus is, on the whole, an

inoffensive chap; and it would, indeed, be a mistake
to suppose that all monkeys are naturally mischievous.
The little Tamarin (*Midas rosalia*) handles its playthings
more carefully than most children, and the females, especially, seem almost afraid to stir without their keeper's
permission. Gratuitous destructiveness is rather a dis-
tinctive trait of the African quadrumana, and their representative in this respect is perhaps the *Cercopithecus
Maurus*, the Moor-monkey, or *monasso*, as they call him
in Spain, a fellow who seems to consecrate his temporal
existence to mischief with an undivided and disinter-
ested devotion. This Maurus and his cousin the rock-
baboon are the terror of the Algerian farmer; but the
baboon contents himself with filling his belly, while the
other tears off twenty ears of corn for one he eats, and
often enters a fig-garden for the exclusive purpose of
stripping the trees of their leaves and unripe fruit. In
captivity he cannot be trusted even with a leather jacket,
and, finding nothing else to spoil, does not hesitate to
exercise his talent upon his younger relatives, to the
detriment of their woolly fur. Still, his intelligence and
restless activity make him a prime favorite with the fun-
loving Spanish sailors, and in the Andalusian seaports
every larger household has a monasso or two,—*monos
de cadena*, "chain-monkeys," as the dealers call them, a
Moor-monkey and a *cadena* being as necessary concomi-
tants in civilized regions as a king and a constitution. A
rupture of the concatenation creates an alarm as if the

chained beast of the Apocalypse had broken loose, and if an unchained monasso gets a five-minutes' chance at a kitchen or a parlor he can be relied upon to commit all the havoc a creature of his strength could possibly execute in five times sixty seconds: an instinct bordering on inspiration seems to tell him at the first glance where and how to perpetrate the greatest amount of actual damage in the shortest possible time. In a harbor-hotel of Cartagena I saw a mono whose terpsichorean talents had made him a more than local celebrity. He could dance the Moorish *zameca*, besides the bolero and fandango, and was sometimes released at the request of his admirers, who pitied his constant collisions with the lock of his drag-chain; but on such occasions the landlady used to charge a *real* extra, for even her presence did not prevent the mono from indulging his ruling passion. Under pretext of returning the caresses of his visitors, he managed to abstract their buttons, upset a flower-pot or two, or interrupted his performances to make a grab at a litter of poodle puppies on the veranda. His scar-covered skull proved that the lot of the transgressor is hard; but the depilated condition of his neck was owing to a peculiar trick of his, as the *posadera* explained it. He would hug a post near his couch under the veranda, and, stretching his head back and his tongue out, would twist his neck to and fro, as if in the agonies of strangulation. During a temporary absence of their mother he once succeeded in deceiving the children by these symptoms of distress: they loosened his chain-strap an inch or two, but happily took the precau-

TOTAL DEPRAVITY.

tion to shut the house-door and the cellar-gate. But they had forgotten the poultry-house; and when the lady returned in the evening her sixteen hens had been converted into Platonic *homunculi*,—" bipeds without feathers and without the power of volition." On another occasion he came near setting the house on fire by drenching the cat with the contents of a large kitchen-lamp. Still, after trying sundry other four-handers, the lady declined to part with her monasso, though she lamented his utter want of principles, like the *Devin du Village:*

> Hélas! que les plus coupables
> Toujours sont les plus aimables!

The *Cercopithecus Maurus* is the near relative of the Indian macaques, beyond any doubt the most interesting pets of their size. Without the pensive despondency of the larger apes, the Cercopithecus Macacus has a large share of their *reasoning capacity:* in whatever way we may choose to explain his intelligence, we certainly cannot ascribe it to instinct. The instinctive faculties of animals are limited in the nature of their purpose,—working in a certain direction with a perfect adaptation of means to end, but narrowly objective,—while the subjective capacity of our four-handed relatives is convertible and pervertible to all possible good, bad, and frivolous purposes: a monkey's mental process subserves the intents of his individual caprice rather than the interests of the species. On my last visit to Antwerp I bought a young Siamese bonnet-macaque (*Macacus radiatus*), whose conduct under circumstances to which no possible ancestral experiences could have furnished any precedent has often convinced me that his intelligence differs from the instinct of the most sagacious dog as essentially as from the routine knack of a cell-building insect. His predilection for a frugal diet equals that of his Buddhistic countrymen, and I have seen him overhaul a large medicine-chest in search of a little vial with tamarind jelly. He remembered the shape of the bottle, for he rejected all the larger and square ones, and after piling the round ones on the floor began to hold them up against the light and subdivide them according to the fluid or pulverous condition of their contents. Having thus reduced the number of the doubtful receptacles to something like a dozen and a half, he proceeded to scrutinize these more closely, and finally selected four, which he managed to uncork by means of his teeth. Number three proved to be the bonanza bottle, and, waiving all precautions in the joy of his discovery, Prince Gautama left the medical miscellanies to their fate and bolted into the next room to enjoy the fruits of his enterprise in his favorite corner. A dog's nose might have saved him all that trouble; but no dog in the world could have devised a plan of simplifying the investigation in default of his physical senses.

Neither a dog nor a monkey is naturally a nest-building animal, and on a cold day a terrier would content himself with crawling into a warm corner; but Buddha has noticed that the sun of my hearth is apt to wane in the eleventh hour, and obviates that contingency by collecting all the loose rags and papers he can lay his hands on whenever the state of the weather threatens a cold night. As a last resort he offers his enemies a truce and bundles in with one of the dogs,—with the poodle generally, on account of his calorific fleece. He was in the habit of utilizing a young spaniel bitch in that way, but toward dark the dog was subject to a fit of whining and scratching and had often to be ejected as a common nuisance,—till Buddha, giving his bedfellow the benefit of his superior foresight, saved her and himself from the discomforts of a cold night's lodging by forcibly suppressing her symptoms of uneasiness. In his intercourse with his two-handed protectors his attachments are not very demonstrative, but his affection, just like a child's, becomes more intimate in moments of real or imagined personal danger. I took him out to a picnic one day, but the festivities were interrupted by the customary thunder-storm, and I was glad to accept a seat in the tent-wagon of one of my next neighbors. Before we got home the rain had swelled the little creek to a torrent, and, finding the ford impassable, we had to make a détour to the next bridge. It was pitch-dark when we reached it, and, hearing the booming of the creek, I jumped out to reconnoitre the safety of the passage. The bridge was in its place yet, so I hallooed to the driver to come on; but through the rush of the water and the rumbling of the coach I heard the uproarious laughter of the occupants. Somehow or other the monkey had noticed my absence and gone almost crazy with excitement. Remonstrances and caresses were quite in vain: he screamed like a madman, and was in the act of jumping out, when I laid hold of him and called him by his name. Recognizing my voice, he flew at my throat, fastened his teeth in my collar, and, thus clasping my neck, gave vent to his feelings in a curious kind of spasmodic sobs. The farmer's girls finally lugged him to the front of the wagon; but every now and then he came back to my corner and tried to establish my identity by passing his hands over my face and feeling for my beard. When his offences against the eighth commandment had roused the wrath of the housekeeper, he used to hide under the stove; but on one such occasion, while the duenna was after him with a broom-stick, a strange dog happened to enter the kitchen, and, without a moment's hesitation, Buddha chose the least of two evils, and, flying into the woman's arms, clung to her for protection, though he had to take a good thrashing into the bargain.

Monkeys are practical physiognomists, and can read half-suppressed emotions in the symbolism of the human face. An angry look at once puts them on their guard. They have an eye for individual dispositions and foibles. During a two years' residence in the suburbs of Vera Cruz I often left Buddha in charge of my landlord, or rather of his children, for the old man was a hipped Cuban refugee and very apt to drown his cares in *aguardiente*. When I came home in the evening, a single look at the monkey-perch told me if the Cubano had been once too often "round the corner," for in that case the Macacus radiatus was hiding behind the curtain or under the sofa, unwilling to meet the enemy in single combat. But the appearance of an ally at once restored his courage: as soon as I entered the room he sallied forth, and seemed to defy the wrath of the tyrant by marching up and down with a strutting gait and an occasional wink at the neutral by-standers. When I pretended to go out and leave him in the lurch, he would sneak along the wall to regain his ambuscade by a roundabout way, and remained as still as a mouse as long as the angry voice of the colonel kept the room under martial law. But if the irate hidalgo selected a scapegoat among his boys, the monkey reappeared in the background as soon as the yells of the victim told him that matters were approaching a crisis, and, taking advantage of the general confusion, would make a raid on the table and fill his cheek-pouches with substantials. After a successful foray of this sort the house-dog often joined him in his retreat, and, instead of resisting his communistic claims, the Macacus then submitted to black-mail, and only now and then silenced the demonstrations of the quadruped by an angry gesture: " Hush up, you fool!" in the plainest language of dumb show. But whenever the obfuscation of the Cubano reached the hypnotic stage, Buddha's tactics underwent a corresponding change: he sallied boldly, mounted the prostrate refugee with a view-halloo whoop, and sometimes proceeded to search his pockets with all the cool effrontery of the *Neveu de Rameau.* His pragmatical speculations on the condition of a top-heavy foe may be rather vague, but he is evidently gratified at the reversion of the order of mental precedence between himself and the big biped. After a flagrant breach of the domestic by-laws he will often forego a couple of meals rather than leave his hiding-place, and only curiosity will bring him out at such times.

In regard to their rank in the scale of intelligence the various quadrumana might be classified after the degree of their curiosity; and I cannot help thinking that man himself owes his supremacy as much to the inquisitiveness as to any moral virtue of his primogenitor. The American sapajous are rather incurious creatures in comparison with their Oriental congeners: no special correspondent in the Divan of the Padisha can be more wide-awake than a macaque in the presence of a stranger or upon his first arrival in a new lodging. Nothing escapes his restless eye: the swaying of an ivy-leaf at the window, the vibration of the teapot-lid, the slightest movement of a strange dog, at once attract his attention and become objects of his vigilant interest. If I am going to refill my mucilage-bottle, I must take care to divert the macaque's attention to the opposite end of the room; when I am sealing a letter, I have to touch sundry other articles on the table, or Buddha will try to find out what I have been hiding in that envelope with such particular care. He had devised a way of opening his cage by sticking his fingers through the bars and lifting the bolt from below; but I baffled his ingenuity by plugging the hinge with a wooden wedge, and the next time I released him he mounted the cage as soon as I turned my back, and began to scrutinize the door with the unmistakable intent of discovering the obstructive innovation. In the first month after his arrival in the United States he was sitting in the chimney-corner with a little Brazilian coaita, when the cold rain suddenly changed into a snow-storm. Both monkeys flew to the window, and, after contemplating the phenomenon in mute surprise for the space of ten or twelve minutes, began to exchange inquiring looks with a peculiar *sotto-voce* chatter, as if the portent had almost taken away their breath. But the conduct of the coaita may have been prompted merely by the example of her elder companion, for she contents herself with enjoying the warmth of the fireplace, while the Asiatic seems to take an abstract interest in the process of combustion, the crackling of the fuel especially, and the occasional eruption of a streak of flaming gases. Not far from my present dwelling-place a suburban railroad company is digging away at a limestone bluff in the way of a projected branch line. The heavier rocks are drilled and fractured with dynamite, and about every six hours a series of detonations go off in quick succession like the shots of a Gatling gun. My menagerie-box is a picture at such moments. The four-handers at once huddle together and accompany each discharge with a convulsive start or a simultaneous attempt to force the door, while the quadrupeds just look up and go to sleep again. When the blasting commenced, an infant capuchin-monkey always became the centre of an excited group : his senior relatives crowded around him with a sudden appearance of eagerness to protect the jabbering brat.

The study of a certain peculiarity in the character of men and monkeys may have induced that shrewd old Theban Epaminondas to reorganize the army of his native state by a division into clans and brotherhoods. The primates of the animal kingdom become heroic in defence and *in the presence* of their friends. If a single boy be caught on the wrong side of an orchard-wall he will give up at once, and generally manage to propitiate the wrath to come by an unconditional capitulation : a pair of chums in the same predicament are almost sure to make matters worse by their defiant sauciness. The same with monkeys: defiance of human authority by means they would never dare in defence of their own lives they will risk for the sake of their companions. It is said that a man can make his own dog bite him ; but the experiment might fail to succeed with several species of monkeys: a macaque, I believe, would rather die than use his teeth in vindication of his private wrongs against the dread chief of the primates; yet this same Macacus will fly like a bull-dog at any man or any number of men who dare to molest his favorite companion. The above-mentioned young capuchin has a full share of the squealing propensities of his species, and if I lay hold of him his outcries never fail to bring Buddha to the rescue. He does not offer to bite me, as long as there is any doubt about my intentions, but in the mean while serves an injunction by grasping my coat-tail and contracting his brows in a menacing way. With other animals this instinct is limited to the protection of their young; though something like a defensive and offensive alliance of friends has been observed among the larger *phocæ*,—sea-bears and sea-lions,—and, strange to say, is not rarely found among *geese;* a single goose is an arrant coward, but a pair of them are liable to become belligerent. The protective association of wild hogs is something quite different,—a sort of *esprit de corps*, founded not on individual friendships, but on the strength-in-unity principle, a courage *en masse*, strictly proportioned to the numerical strength of the confederation.

The language of our little cousins has a sound or a gesture for every emotion. In his fits of loving-kindness the black spider-monkey chirps like a bird, and hugs the objects of his affection with such a fervor of kindness that dogs and cats have often to use their teeth to escape suffocation. In a huff he struts up and down with his long tail straight erect like an Hungarian pike-standard. The pretty vevet (*Cercopithecus callithrix*) has an amusing way of intimating his desire for food by moving his head to and fro with alternate simpers and grins. In a fit of anger he gets his back up like a panther crouching for a spring, and claws the floor as if he were scratching or tearing something. Prince Gautama

UNREQUITED LOVE.

expresses his displeasure by a sort of smacking click and a chattering movement of his jaws. If he is petted or wrapped up in a shawl on a cold morning, he protrudes his lips with a quite peculiar mumbling purr more nearly resembling a certain modulation of the human voice than any animal sound I am acquainted with. His signal of alarm is a coughing scream, not unlike the yell of a frightened dog. The meaning of that scream seems, indeed, to be understood by every beast or bird, as certain onomatopoetic words recur in the language of every nation. The screech of the capuchin-monkey is somewhat louder and shriller: an adult of the white-faced variety, a fellow not much larger than a cat, can out-yell a couple of good-sized boys. Nearly all the South-American ring-tails are obstreperous brutes, and their talent culminates in the big red howler (*Mycetes ursinus*), a vocalist whose performances, combined with the screams of the jaguar, make the nocturnal forests of the Orinoco a howling wilderness in the most shocking sense of the words. The meaning of his nightly uproar is rather doubtful, since it can hardly be a love-note, like the amorous acclaims of the red deer and buffalo at certain seasons of the year. It may be intended to frighten his enemies; and if it answers that purpose a troop of Mycetes cannot complain of want of elbow-room, for the whoops of the old sachems can be plainly heard at a distance of four English miles. Besides this astonishing vocal power, the chacma baboon is a still greater master of the science of *tucbeer*, the stentorian art of intimidating an enemy, so much valued among the ancient Saracens and modern Sioux. The hoarse, coughing bark of the male chacma expresses, indeed, a very paroxysm of savage passion, and, added to his ferocious appearance in a fit of rage, may well frighten the Namaqua nymphs out of their scanty wits.

The anthropoid apes are a somewhat taciturn race, but a chimpanzee's murmur of affection is very expressive, and quite different from his grunt of discontent. A sick orang-outang sheds tears, moans piteously, or cries like a pettish child; but such symptoms are rather deceptive, for the orang as well as the chimpanzee is a great mimic, not of men only, but of passions and pathological conditions. Two years ago I took temporary charge of a young chimpanzee who was awaiting shipment to the Pacific coast. His former landlord seemed to have indulged him in a penchant for rummaging boxes and coffers, for whenever I attempted to circumscribe the limits of that pastime my boarder tried to bring down the house, metaphorically and literally, by throwing himself upon the floor and tugging violently at the curtains and bell-ropes. If that failed to soften my heart, Pansy became sick. With groans and sobs he would lie down in a corner, preparing to shed the mortal coil, and adjusting the pathos of the closing scene to the degree of my obstinacy. One day he had set his heart upon exploring the letter-department of my chest of drawers, and, after driving him off several times, I locked the door and pocketed the key. Pansy did not suspect the full meaning of my act till he had pulled at the knobs and squinted through the keyhole, but when he realized the truth life ceased to be worth living: he collapsed at once, and had hardly strength enough left to drag himself to the stove. There he lay, bemoaning his untimely fate, and stretching his legs as if the *rigor mortis* had already overcome his lower extremities. Ten minutes later his supper was brought in, and I directed the boy to leave the basket behind the stove, in full sight of my guest. But Pansy's eyes assumed a far-off expression: earth had lost its charms: the inhumanity of man to man had made him sick of this vale of tears. Meaning to try him, I accompanied the boy to the staircase, and the victim of my cruelty gave me a parting look of intense reproach as I left the room. But, stealing back on tiptoe, we managed to come upon him unawares, and Pansy looked rather sheepish when we caught him in the act of enjoying an excellent meal.

Jules Michelet asserts that few women know any medium between love and hatred; but his paradox is strictly true in regard to the sympathies and antipathies of some of our Darwinian relatives. In a domesticated ape's intercourse with strangers open hostility is generally the only alternative of importunate endearments. The pros and cons are decided very promptly by some inscrutable criterion, for his loves at first sight are by no means biassed by prepossessing appearances nor even by friendly overtures on the part of the biped. It may be something more than caprice: a monkey may detect the insincerity of a caress or good nature under a gruff mask by symptoms that escape the human eye. Still, their conduct toward visitors depends somewhat upon circumstances : in private interviews they treat strangers with a cautious reserve, often evidently suggested by the idea of having been surrendered into the hands of an enemy, possibly of a new master whose good will it might be advisable to conciliate.

Humility is said to be the virtue of those who have no other merit. Judging from the analogies of the brute creation, the rule would seem to admit of occasional exceptions: some very meek and lowly animals are by no means without parts; but the most gifted of all, the doubly-ambidextrous four-hander, is certainly the most self-asserting. In a congregation of miscellaneous mammals a monkey at once assumes command; he knows his rank, and no Arabian sheikh in an assembly of African chieftains is swifter in resenting any disrespect to the *Cæsarea majestas* of his mental superiority. A female Javanese Manki, the smallest variety of the genus *Cercopithecus*, managed for several months to lord it over the occupants of my zoo-box, till her supremacy was disputed by a Scotch terrier, who had to maintain a national dignity of his own and declined to accept a passive *rôle* in a game of leap-frog. But he soon ascertained that a practical joke can be a lesser evil : the peace of his life was gone; he could not eat, drink, or romp without awakening the anathemas of the Manki, who seemed to watch him like a lynx, and at his least sign of playfulness pretended to fly into a violent passion, chattering and screaming till the quadruped subsided like the victim of a vociferous shrew. His native pride upheld him for three or four days, but before the end of the week he was glad to conclude an armistice on the Manki's own terms, and submitted to anything that would secure him the privilege of eating his meals in peace.

Most monkeys are masters of the art of bullying their fellow-beings by intimidating gestures, especially a sudden erection of the scalp-bristles, combined with the attitude of a bull-dog crouching for a spring. In reality the average small monkey is a poor fighter: his finger-nails are blunt, and his teeth frugivorous and short; but dogs, as well as cats and raccoons, are generally imposed upon by his impudence, apparently concluding that so much assurance must be backed by some occult martial resources. But even upon their retreat before an undoubted physical superior the little half-men yield only under protest: when the Mexican bush-panther continues his forays after daylight, the capuchin-monkeys keep up an incessant chattering, jumping to and fro, as if they defied him to a climbing-match through the tree-tops. Should he happen to catch one of them, the rest will risk their own necks rather than forego the satisfaction of pursuing the murderer hour after hour with furious screams.

No monkey submits without "back-talk:" my Buddha, the tamest macaque I ever saw, would bristle up like a fighting-cock if I thwarted him in his caprices; at sight of a stick he retreated just out of reach, then, suddenly turning, often gave me a bit of his mind, with a coughing grunt and a look like Faust defying the Demiurgus: "Ich bin's, bin Faust, bin deines Gleichen!" His endearments could not be spurned with impunity; men and beast had to choose between his caresses and his wrath; in his younger days especially, he claimed a constitutional right to be petted, and would not stand any slight; the *spretæ injuria formæ* generally threw him into a squealing-fit, and often into such a huff that nothing short of abject flatteries would restore his good humor. Nor is it easy to frighten a delinquent monkey into an unconditional surrender as long as he can elude your grasp; he is apt to dispute the competence of the court, and has to be arraigned by strategy: salvation by flight seems to be a fixed idea of the simian mind. To run down an ape of the larger species is, indeed, no child's play, even in the open fields and under circumstances that would insure the capture of any other terrestrial animal. During the Dutch expedition against Acheen, Captain Hess, of the Batavian Rifles, procured the skin of an old orang-outang who had been chased a whole day by a troop of natives with clubs and dogs and had fairly exhausted their patience before they could get hold of him. They had surprised him on a high *patina*,

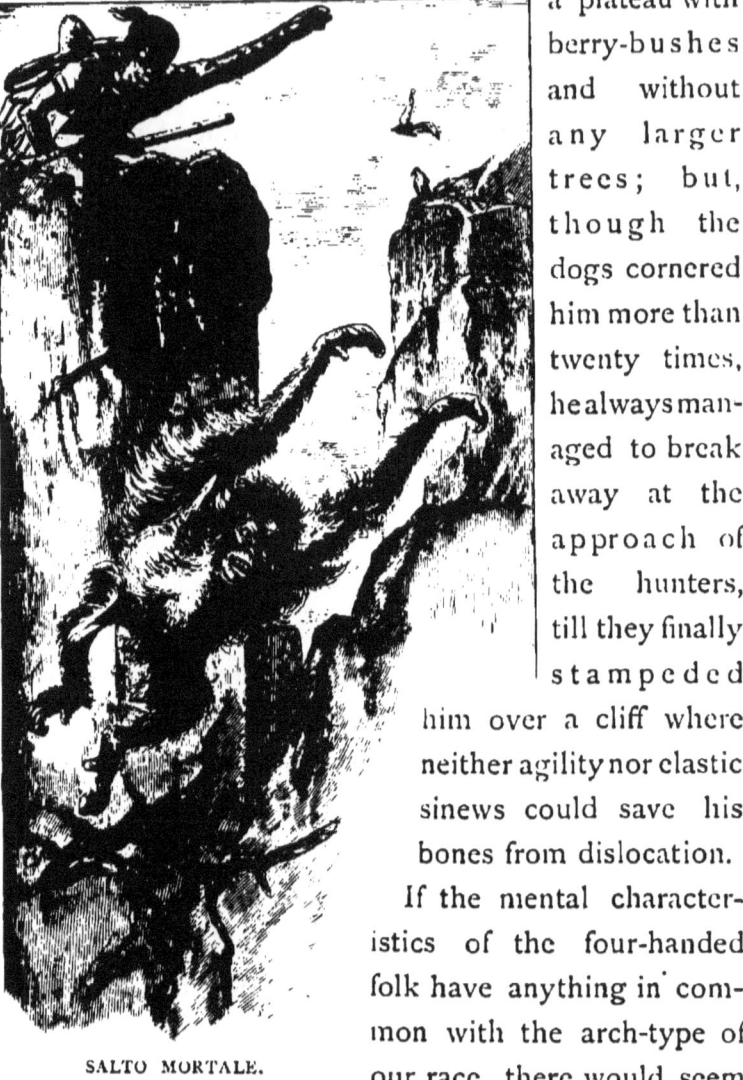

SALTO MORTALE.

a plateau with berry-bushes and without any larger trees; but, though the dogs cornered him more than twenty times, he always managed to break away at the approach of the hunters, till they finally stampeded him over a cliff where neither agility nor elastic sinews could save his bones from dislocation.

If the mental characteristics of the four-handed folk have anything in common with the arch-type of our race, there would seem to be a strong presumptive evidence that man is not *per naturam* a law-abiding animal. The Psalmist may have stretched his poetic license when he assured us that all

men are liars, but there is no doubt that all monkeys
are thieves. They all steal ; in their native land depre-
dation forms their daily and constant employment, and
Nature has done her utmost to equip them for their
trade. The monkey Hanuman is the Indian Mercury,
the patron saint of shoplifters and freebooters. Light-
fingered, quick, and prehensile, his four hands seem especially adapted for pillage ; he carries a double "kit," a
pair of capacious cheek-pouches, for storing his plun-
der; his arboreal domicile furnishes him ready-made
ambuscades and lurking-places; he cannot be caught
napping, his head moves on a versatile fulcrum, and
his furtive eyes are ever on the alert. It is wholly impossible to cure a monkey of his raptorial *penchant:*
you may tame him till he makes your lap his favorite
hiding-place, you may surfeit him with tidbits, but the
moment you turn your back he will ransack the room
from top to bottom and cram his pouches with everything bearing the faintest resemblance to comestibles.
In Hindostan monkeys enjoy all the privileges of a Mohammedan lunatic, being permitted to rob the orchards
with impunity, decimate the rice-crop, and rob all the
birds'-nests they want; but, not content with levying
out-door contributions, they pillage the cottages of the
natives while the proprietors are at work in the fields ;
nay, they often manage to despoil the larder of the foreign
residents, or blackmail their children if they leave the
bungalow with a lunch-basket or a pocketful of nuts.

The Rev. George Thielmann, of the Moravian Mission, who passed several years in the Eastern Punjaub, describes the despair of his German cook at the impudence

MISPLACED CONFIDENCE.

of the light-fingered gentry. "I do not see how the natives can stand it," said she: "if they take those baboons for Christians, they ought to have a penitentiary in every village." If she went to the door to answer a bell, the macaques entered the kitchen through the rear window; going to look after her sun-dried peaches, she found that the Bhunder apes had been beforehand with her; and if she left her bedroom window open she was awakened by a committee of Hanumans taking an inventory of her wardrobe. One day she left the gardener's dinner under a tree where he used to take his siesta, but, returning with a dessert of German doughnuts, she was just in time to see a troop of Rhesus baboons running off with the dishes and bottles.

From the moment that a young monkey is weaned he has to steal, for Dr. Brehm's observation applies strictly and literally to every species of quadrumana: the mother-monkey robs her own child, and forces it to eat its food by stealth. The proprietor of the "Zoological Coffee-Garden," in Savannah, Georgia, has been very successful in rearing young monkeys, and the visitors of his happy-family department can witness the same scene thrice a day,—a number of half-grown capuchin babies fleeing from the wrath of their own parents. As soon as the dinner-bucket is brought in, the youngsters hide in the corner and watch their opportunity, for while their seniors are feeding there is no hope of a crumb or a drop of milk; but sooner or later the old ones are sure to fall out, and during a general scrimmage for a tidbit the children sometimes get a chance at the bucket, and take care to make the best of it. But woe unto them if their progenitors catch them *in flagranti!* Sires, mothers, and aunts combine to avenge the sacrilege, and the noise of the punishment often sets the whole menagerie agog. I have seen a she-macaque jamming her bantling up against the wall and extracting from its cheek-pouches the gifts of a charitable visitor, together with all the crumbs and scraps the little one had gleaned from the floor, and then adding outrage to injury by cuffing the victim's ears.

As a consequence of such treatment, a baby-monkey in the teens of its months is generally as lean as a rake; but the apparent cruelty of its parents may be a wise provision of nature. After Jean Jacques Rousseau's plan, it would be the best possible education for a creature that has to make his living by *stealth*. Hunger sharpens even a baby's wits, and a young four-hander of ten months is really as preternaturally wide-awake as a ten-year-old *gamin* of the Quartier des Savoyards. Having learned to mistrust his own parents, he is naturally very circumspect in his dealings with his human guardians, and after a year of the kindest treatment a mere contraction of your eyebrows is sufficient to drive him grinning and chattering to his hiding-place. A monkey rarely takes an offered present without watching your eyes and then snatching it with a sudden grab, apparently unable to realize your generous caprice, but concluding to take luck by the forelock before you change your mind. In the summer season I have often permitted a tame monkey to run at large, and before the end of the week I invariably found that my roommate had established a *cache*, a hiding-place for storing his mammon of unrighteousness,—stolen apples, nuts, pencils, corks, and often also pieces of meat, or eggs, that became offensive in the course of time and thus betrayed his depository. Buddha, too, was incorrigibly addicted to this kind of nest-hiding, though so fully aware of the illegality of the practice that he took to his heels as soon as I discovered his swag. I once, to try him, put a scalding-hot egg on the table, and went out to watch him through a key-hole. A rumbling in the corner told me that he had descended from his perch, and soon after his head appeared on the farther side of the table. He touched the egg, gave a grin at the door, and seemed on the point of retreating, but, drawing himself up once more, he cast a hurried glance over the table and snatched the stopper of a vinegar-flask rather than return empty-handed. At that moment I opened the door, but Buddha had disappeared, evidently into his usual hiding-place behind the lounge. He knew I had watched him, and thought it prudent to keep out of sight till time or new events had obliterated the memory of his crime.

The English word stalwart is derived from *stael-worth*, —*i.e.*, worth stealing; and the same criterion seems to be a monkey's standard for the value of earthly things in general. Any novel, movable, and portable object at once excites his interest. If the digestible qualities of the novelty seem doubtful, he appears to act on the principle that in the mean while it can do no harm to appropriate it. North of the Rio Grande most capuchin-monkeys are martyrs to rheumatism, and three poor cripples of the *Cebidæ* species had been assigned winter-quarters in the kitchen of a New-Orleans boarding-house. They could be trusted, as their complex ailments disqualified them from running and climbing, their only mode of progression being a sidelong wriggling on their haunches and elbows. But one day the landlady heard a frightful caterwauling, and, entering the kitchen in haste, was surprised to see one of her patients on top

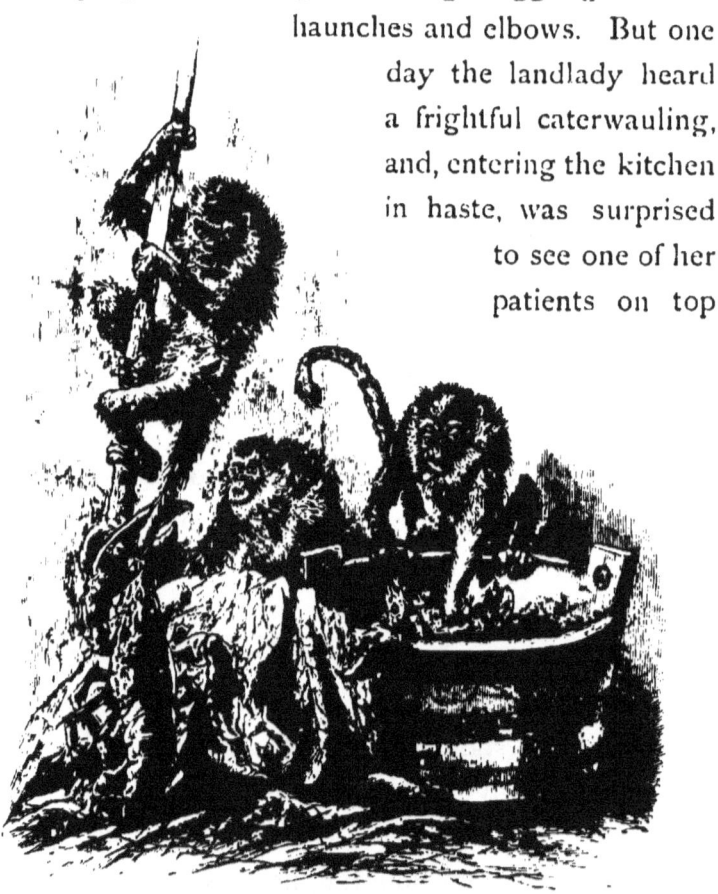

MARTYRS TO FREE INQUIRY.

of the chimney-ladder, while another was rolling about in a fit of fantastic contortions. The cook had left on the floor a bucketful of Pontchartrain crabs, and during her momentary absence the monkeys had fallen victims to the cause of free inquiry. Somehow or other, the cook's manœuvres had drawn their attention to the bucket, and, having managed to upset it, their ring-tails had got entangled with the not less prehensile crustaceans.

It is a curious fact that all the larger varieties of the American monkeys are endowed with a voice of almost superhuman power, while the apes of the Old World are a comparatively silent race. The reason may be that the jaguar, the chief enemy of the Brazilian ring-tails, does not like to be yelled at, while the same expedient would be unavailing against the antagonists, or rather rivals, of the Oriental four-handers. The Rhesus baboon, though one of the demi-gods of the Hindoo pantheon, is about the most undesirable pet a menagerie could get hold of. A big brute of this species, an almost hairless old male, named Bhunder-Beg or Sahib-Onki-Walla, was presented to the new Zoological Garden of Marseilles, but, in spite of his unusual size, the superintendent sent him back by the very next steamer. "I should like to keep him for your kind intentions' sake," he wrote to the donor, "but it won't do: with the sole exception of Petronius Arbiter, your Bh., surnamed S. O. W., is the most obscene personage known to ancient or modern history."

Monkeys are *amical* rather than gregarious creatures :

they like to consort with one mate, one favorite companion, but dislike a crowd. Their larger assemblies have always a special purpose,—a combined attack upon some beast of prey, a foray upon an orchard where outposts are needed; but that purpose attained, the troop separates pair-wise, even in captivity, unless a low temperature should oblige them to huddle together. Wolves, too, as well as many species of migratory birds, congregate only in particular emergencies, while hogs and horned cattle always prefer to herd in the largest possible numbers. The gastronomic predilections of the four-handed freebooters are more uniform than might be supposed from the dissimilarity of their habitats. Sweet or sub-acid tree-fruits always form the staple of their diet; *faute de mieux*, they manage to rough it on roots, nuts, mollusks, and even insects, which, besides a few berries, constitute the only sustenance of the Gibraltar macaques; but *meat*—*i.e.*, the flesh of mammals and birds—seems as repulsive as poison to all daylight monkeys, as well as to the plurality of the African lemurs. But even the *Simiadæ* proper, the Asiatic monos and anthropoids, whom no starvation can drive to carnivorous shifts, are ravenously fond of milk and eggs, thus justifying the theory of the "Liberal Vegetarians," who distinguish between animal and semi-animal articles of food.

Most monkeys are gourmands, and their alleged fondness for stimulants is a favorite argument with the opponents of teetotalism; but the truth is that bibulous monkeys, like boy topers, owe their *penchant* to the corrupting influence of their associates rather than to an innate tendency. Ninety-nine per cent. of our menagerie monkeys have crossed the Gulf of Mexico, if not the ocean, and Jack Tar would rather forego his own tipple than miss the fun of forcing grog or tobacco upon his four-handed passengers. That they contract a passion for such things proves not their but the ancient mariner's natural depravity, and that they indulge the habit with temporary impunity demonstrates only the marvellous faculty of adaptation which the quadrumana share with their two-handed cousins. It is true that wild apes are sometimes caught by means of intoxicating baits; but for such purposes the taste of the alcohol has to be disguised by a liberal admixture of saccharine elements, and I would wager any odds that a new-caught monkey would prefer the sourest crab-apple to a piece of the best Schweizer-käse or chewing-tobacco. No danger will deter a monkey from grand larceny if he gets a chance at a good store of candies or preserves; it must be seen to be believed in what a short time a little macaque will put himself outside of a boxful of sugar-plums. Sugar is his Paracelsian quintessence, the elixir of life and joy; and I suspect that in pursuit of that *summum bonum* he will swallow considerable quantities of *per se* hateful fluids, just as many a juvenile coffee-drinker would prefer his sweetening "straight." Salt, on the other hand, is his grand aversion, and, for all we know to the contrary, Sylvester Graham may be right, that only our carnivorous habits oblige us to swallow a daily dose of chloride of sodium.

But, if the views of Luigi Cornaro are correct, there is no doubt that the brute creation must be tainted with original sin : Jacko is no friend of homœopathic rations; of such comestibles as Nature has intended for him he wants to eat his fill, and quite literally, too; the Colobi and Cercopithecs actually devour the utmost amount of food compatible with the calibre of their digestive organs; the slender egret monkey (*Cercocebus Aygula*), for instance, eats with ease a daily quantum exceeding four-fifths of his own weight. In the Zoological Garden of Schoenbrunn, near Vienna, I saw one that would never refuse a tidbit : after stuffing himself with apples, crackers, and untold cherries, he still contrived to find room for a large piece of ginger-cake. *Fruges consumere nati*, they act their part well : if their appetite increases with their size, a troop of sacred Hanumans must severely strain the tolerance of the Brahminical natives. An anthropoid ape has the stomach of an Arkansas tramp and the lungs of a hectic school-girl.

Few orangs or gibbons outlive the third year of their captivity; the least defect in the ventilation of their prison amounts to a death-warrant; every winter month seems to shorten the term of their life by a year or two, for in the tropics their average longevity exceeds a quarter of a century. The pulmonary diseases of the human species have less to do with a low temperature than with the impurities of the in-door atmosphere, and the effluvium of a menagerie is notoriously offensive: still, it is a strange fact that small monkeys, like squirrels, can for a long time subsist on a very minimum of life-air. I have seen my macaques crawl into a pile of cast-off clothing and cover themselves, head and all, with a sixfold stratum of coats and blankets before going to sleep; and during the coldest nights of an Ohio winter a dealer in zoological sundries kept a spider-monkey alive by bundling him up with a couple of fluffy terriers. His method was to chuck them into a sack half full of wool and hay, tie the sack, put it into a barrel, and cover it with an extra blanket or two, according to the state of the weather. He has tried the same plan with squirrel-monkeys and capuchins, and his success seems to prove that animals can get along with less air than is generally supposed. The nest of the common ground-squirrel is even a greater puzzle to the zoological physiologist: long before the beginning of the hibernating season the little bobtails retire to the bottom of a hole where all the air that can possibly reach them has to penetrate an eighteen-inch mass of compact moss and hay, often besides a thick layer of dead leaves and rubbish.

A somewhat paradoxical character-trait of the more intelligent four-handers is their antipathy to children.

The gibbons, baboons, Bhunder monkeys, and all the larger macaques fly into a passion at the mere sight of a young biped. The much-plagued menagerie apes might plead a legitimate reason for this aversion; but the same peculiarity may be observed in monkeys that never had an opportunity to make the acquaintance of a French cabin-boy or American Sunday-school excursionist. It may be jealousy, an envious rancor against the natural competitors for the affection of their master,—akin to a lap-dog's malice toward a rival pet; or else it is perhaps a manifestation of a secret antipathy to the *gens humana* in general,—a misanthropical *penchant* restrained in the presence of the dread king-ape, but exploding against the saucy princes, as a man might be tempted to suppress a young Cyclops before his attainment of a dangerous age. A monkey will risk a good deal for the fun of teasing an *homunculus*. I never saw an old macaque miss an opportunity of that sort. Buddha, especially, was the terror of my young visitors. In a crowd of half-grown boys he contented himself with defiant gestures and a volley of chattering imprecations, but with youngsters under five he at once proceeded to active hostilities, pulling their ears or biting them in a way that could not be mistaken for a practical joke. The son of a German colonist in New Freiburg, Brazil, was once attacked by a swarm of Mycetes monkeys whom he had noways offended; and an English traveller mentions a case of a little child being killed by a troop of Ceylon wanderoos:

"A flock of these animals may be seen frequently congregated on the roof of a native hut; and some years ago the child of a European clergyman stationed at Tillipally, having been left on the ground by the nurse, was so teased and bitten by them as to cause its death." (Sir Emerson Tennent's "Ceylon," vol. i. p. 132.)

Monkeys seem to believe in the efficacy of vicarious atonement. A little yellow bitch, whose couch my pet macaque has shared for the last two years, has to suffer grievous indignities as his monkeyship's scapegoat. Whenever I detect him in any misdeed, he turns upon his partner with a look of severe disapprobation, and if his iniquities bring him to grief he "takes it out" of her, with a promptitude that has taught her to take to her heels as often as I arraign him for an unpardonable offence, and sometimes even during the perpetration of his sins.

As exemplar of the virtues often attributed to a state of nature, monkeys are, indeed, rather an indifferent success. Their standing in the peculiar graces of self-denial and self-abasement is certainly below the Christian standard; but, for all that, one cannot help observing the ways and tricks of the little sinners with an interest entirely distinct from the pleasure attached to the study of natural history in general. Can it be something more than the mere scientific curiosity of the professional zoologist? The question is perhaps answered by Arthur Schopenhauer's definition of a representative monkey: "An epitome of man without the human faculty of dissimulation."

CHAPTER II.

MOUNTAIN SHEEP.

IT is wonderful under what difficulties some wild animals have managed to survive the endless warfare of man against nature. Only island-dwellers have succeeded in utterly exterminating any species of their fellow-creatures. The dodo of the Mauritius, the blue parrot of the Norfolk Archipelago, and the Newfoundland auk (*Alca impennis*) lived and perished within their respective island-homes; the New Zealand *moa*, too, is supposed to have become extinct in recent ages,—*supposed*, I say, for it is by no means certain that the gigantic bones discovered by Tasman and Hochstetter were not of antediluvian origin. But on the mainland even the large mammals have thus far successfully maintained the struggle for existence. Danger has sharpened their protective instincts, and, by a wise law of Nature, the very scarcity of an animal race improves the life-chances of its surviving representatives. The coyest female will encourage the suit of the last male of her species, reduced food-stores may still supply the wants of a reduced number of consumers, and, above all, persecution abates when there is little left to persecute: the most ruthless and indefatigable of hunters will hardly care to track and run down the last band of Norwegian reindeer or the last pair of African gorillas.

For the same reason, I do not believe that the wild sheep of the North American continent will ever entirely disappear from its mountain-haunts. The mountain sheep or cimarrón (*Ovis montana*) has many enemies and is not very swift-footed, but it is probably the shyest quadruped of the New World. On the treeless highlands of our Central States it is no easy matter to get within rifle-shot of a herd of "bighorns," as the Colorado trapper calls them, but on their favorite pasture-grounds in the Pinos Altos range, in Southern New Mexico, the prospecting miner can sometimes approach them at the time when the wild-rose-bushes are in full bloom and confound the scent of the wary outposts. A herd of grazing cimarróns is a curious sight: they do not content themselves with posting a single sentinel, after the manner of the antelopes and wild llamas, but all the veterans, especially the nursing ewes, take their turn at the picket-post, and every now and then run to the next rock and rise on their hind-legs in order to enlarge their field of view. A low snort, accompanied by a stamping or scraping kick, is a sign of vague suspicion, and puts the whole herd on the *qui vive;* even the young kids crowd around their dams and anxiously await the next word of command. The sudden side-leap of an outpost is a signal of imminent danger; like a well-drilled squadron the herd at once wheels around and gallops away in a direction which the leaders seem to have precalculated for every possible emergency. During their winter migrations from sierra to sierra the sachems of a large herd become as cautious as the leaders of the Anabasis, and will often stand immovable for hours together at the brink of a plateau, with their eyes fixed upon some doubtful object in the neighborhood of their meditated line of march. If the outlook is not quite satisfactory, they decline to take the benefit of the doubt, and stick to their vantage-ground till the coast is decidedly clear. In the winter of 1874 a company of American engineers put up a line of telegraphs from Matamoras to Saltillo, in Northern Mexico, and on four consecutive days they saw a number of cimarróns approaching their camp from the direction of the San Cristoval Mountains and retreating again like the scouts of a circumspect guerilla leader. But on the following Sunday a large herd crossed the road, heading due south toward the Sierra Mesilla, in Western Durango. The continual extension of the wire line and the noise of the workmen had delayed their march, the pilgrims having evidently bided their time in what the French call a "camp of observation."

The Mexicans assert that the mountain sheep never stays within earshot of a permanent human settlement, and that the cimarrón population of their border-states has been considerably increased by emigrants from the North. There is no doubt that the freedom-loving monteros have steadily retreated before the advance of our noisy civilization,—first westward, and lately both southward and northward, from the neighborhood of the great trans-continental highway. Colonel Pennypacker, of the United States army, has told me that he remembers the time when the "bighorns" were as abundant as mountain quail in Western Colorado, and that the officers of Fort Garland used to kill them by dozens in the vicinity of the fort. They are now found only near the head-waters of the Gunnison River; and if the Leadville Railroad should be extended to the Colorado Valley they will probably leave the State altogether. They have already left Nebraska, Utah, and Southern Wyoming, and even in the northern part of the territory the name of the "Bighorn Mountains" is fast becoming an anachronism. In the Southwest they have maintained their ground much better (carnero-meat is a drug in the markets of Chihuahua, Tucson, and Santa Fé, and they are still pretty abundant in the Sierra Nevada of Southern California), but also in the far Northwest—for their southward migration has nothing to do with climatic predilections: the mountain sheep is as hardy as the grizzly bear. The Montana prospectors meet great herds of them in the main chain of the Rocky Mountains, but especially in the icy summit-regions of the

Pend d'Oreille range, on the borders of British North America. Even in midwinter they shun the valley settlements. During ice-storms that drive the black bear to his den and kill black cattle in the river-valleys the cimarrón survives where the hill-foxes wander, in the pine wolds and box-elder coppices of the dreary uplands.

WINTER QUARTERS.

In November, and sometimes at Christmas, the miners of Bannock City hear the cry of the old rams in the highland-gorges,—a long-drawn, booming bark; not a signal of distress, but an amatory acclaim, an invocation of the *dulcis Dea Amathusia* when the mercury trembles at forty-five below zero. In stress of weather the cimarróns generally take refuge in a lee-side pine grove, and are thus sometimes cut off from their pasture-grounds, snow-bound, for a month or two, and have to rough it on pine sprouts and such roots and herbs as they can scrape up in the deep-frozen mould.

A party of Mormons, being caught in a snow-storm while crossing the Wahsatch Mountain in 1849, were saved, according to Elder Millard's report, by coming across a sheltered cove in the piny woods where a troop of mountain sheep had trodden down the snow and cropped the branches as high as they could reach, thus forming a series of snug pine arbors,—a ready-made tabernacle for the necessitous saints. In this instinct of finding shelter-places from the cold mammals are far superior to birds, probably because they cannot emigrate so easily. On the bitter-cold New-Year's morning of 1871 the game-keeper of the Duke of Gotha picked up not less than thirty score of dead crows in his master's rookery at Rheinhards-Brunn, but a band of fallow-deer had saved themselves by breaking the lath door of a cellar-like grotto and crowding into the innermost corner of the vault. Besides, I believe that most wild beasts have a little of that talent for hibernation which helps squirrels and badgers over the worst hours of the long *Biornir-nott*,—the "bears' night,"—as the old Germans called the winter season. During a heavy "norther" buffaloes often stand in the hollows of the Texas crosstimber for days together in a semi-torpid state, and the little musk-ox must probably draw considerably upon his inner resources to survive the terrible snows of the Hudson's Bay territory. It is also certain that some quadrupeds, including the mountain sheep and the guanaco, are able to distinguish the signs of an approaching storm from those of a common thunder-shower. Mexican shepherds have often been warned to save their flocks by the mad gallop of a troop of mountain sheep fleeing toward some sheltered valley on the lee-side of a wind which gradually rose to a destructive hurricane.

Frederick Gerstaecker found a cimarrón camp on the very ridge of the Sierra Nevada, but no hunter, so far as I know, has ever discovered the lying-in establishment of a mother-ewe; the cimarrona seems to summon all her secretiveness and topographical experience to hide her new-born lambs from human sight. In August or late in July—rarely sooner—they are found in company of their seniors, evidently numbering their days by weeks, but still rather misshapen, chub-headed, and ridiculously long-legged little fellows, resembling fallow fawns rather than lambs. The whole family, indeed, has something cervine in its appearance. Nature is said to abhor a vacuum, but shows a still more decided repugnance to systematism, and seems to take a special delight in puzzling our zoological categorists.

There are animals that refuse to be classified. The Swiss nuthatch (*Sitta europæa*) is, in habits and appearance, half titmouse and half woodpecker; the South African proteles looks like a hybrid between a civet-cat and an hyena; and the Rocky Mountain sheep holds the exact middle between a sheep and a deer. In the formation of his neck, head, and horns he resembles the Sardinian moufflon-wether, but his rump, stump tail, and legs are those of the Virginia deer; his color, too, is a brownish dun, and his hair is straight and short, with the exception of a wreath of long bristles at the base of the neck. The lambs are whity-brown, with the same dark streak along the spine that is sometimes seen on fawns and very young colts. A *fox-squirrel* the *Sciurus cinereus* is called in the Southern Alleghanies: *deer-sheep* would be the most appropriate English name for the carnero cimarrón.

On a sudden stampede young lambs often get separated from their dams, and have sometimes been taken alive. They can be brought up with the kids of a milch-goat, and get tame enough to follow their foster-mother to the valley, though they prefer the south side of a hedge to the most comfortable stable. Domesticated rams are apt to be troublesome, for an old cimarrón is as irascible as a fighting bull, and has a disagreeable way of charging his adversary from behind,—not rearing and plunging like a billy-goat, but running full tilt and with an unmistakable business-purpose. If permitted to roam at large, he is given to solitary rambles among the cliffs, and is liable to lose his way if he has once ascertained the difference between coarse prairie-grass and the aromatic herbage of the upland leas. But, like other savages, the cimarrón can be subdued by his vices. The craving of his ruminant stomach for salt easily degenerates into a fondness for stronger stimulants,— tobacco, cider, and *aguardiente:* in quest of a "chew" he will besiege his master's door and button-hole strangers with the persistency of a begging friar. Tippling, however, does not improve his temper: the most petulant pet I ever saw was the wether Panchito, a domesticated cimarrón of such intemperate habits that he was repeatedly expelled by his first owner, who at last presented him to the sexton of the Chihuahua cathedral. I came to Chihuahua in 1873, and was delayed almost forty-eight hours by the failure of the stage-driver to procure relays; the festival of Santa Maria de Guadalupe had set the city agog, and all horses and mules were strutting in the cavalcade procession bedecked with flags and orange-rosaries. On the afternoon of the second day the festival came to a crisis; the doors of the cathedral were thrown open, and the votaries surged in and out, helped to drag the ecclesiastic howitzer to the centre of the plaza, and crowded around the open-air pulque-shops. The national drink flowed in streams: *pulque*, a Mexican will tell you, does not induce drunkenness, but only *borracheria*, a mild form of obfuscation less inconsistent with the character of a Christian and a gentleman. The white caballeros certainly managed to keep their heads level, and the ragged mestizo lying on his belly in front of the esplanade had only lost the use of his legs, since the activity of his consciousness asserted itself by a triumphant yell whenever the howitzer was fired. At the third shot, Don Panchito bolted from the basement, himself evidently laboring under an incipient stage of *borracheria*, for at the next discharge he jumped up with all four legs at once, and then, spying the yelling Indian, made a rush and "fetched him one in the ribs," to the uproarious delight of the assembled Chinacos. Sixteen more shots were fired, and sixteen times Panchito charged the Indian, whom he somehow seemed to connect with the cause of the obstreperous demonstrations. He then turned his attention to the school-girls, whose long scarfs appeared to excite his disapprobation, and was going to tackle a young lady with a conspicuous shawl, when a well-aimed kick from her gallant sent him spinning into the basement-vault. But just before I left he reappeared, like Satan *ex infernis*, and when I saw him last he was butting the choir-boys as they sallied successively from a side-porch.

Domestic sheep that lose their way in the sierra are sometimes butted to death by the wild bighorns; but this cruelty is inspired less by malice than by that singular instinct which impels gregarious animals in a state of nature to destroy the decrepit members of their tribe. The cimarrón, recognizing the *Ovis domestica* as his near relative, is scandalized at her fatness, stupidity, and helplessness, and possibly considers it his duty to put her "out of her misery." But, if he does not spare his poor kindred, he certainly does not spare himself either. Frederick the Great's dictum seems to be his motto: "*Il faut traiter son corps en canaille.*" In merciless winter storms he will fly against the wind at a tearing gallop for hour after hour, and he rarely descends from the highlands on account of the weather only. Wounded to death, he still tries to keep up with his flying companions; I have seen a young ram struggling to his feet again and again with a load of buckshot in his lungs, stamping the ground impatiently at his growing weakness, till he finally fell over on his side, almost *exsanguis*, but working his hoofs to the last. The cimarrón cannot be "cornered," like the Swiss chamois, surrounded, and captured at the edge of a precipice; driven to such extremes, the leading ram leaps down into certain death, and the herd will follow unless they are numerous enough to break the blockade with the chances in favor of a few survivals. Declivities of twenty or thirty feet will not stop them: they have a wonderful knack of alighting on their hoofs. There is a prevalent notion that mountain sheep in jumping from a high cliff will alight on their horns; but that is a mistake: they jump off head foremost in order to keep their bal-

MOUNTAIN SHEEP.

A STEEP ALTERNATIVE.

ance, but, on approaching the ground, take care to save their lives by stretching out their fore-feet in the nick of time. De Mora, in his "History of Mexico," goes so far as to assert that the carnero cimarrón cannot be killed at all by a fall "unless he should happen to drop on the sharp peak of a rock." A bighorn ram attains a weight of from one hundred and fifty to two hundred pounds, and exceeds the domestic sheep in size, and I am sure that a plump fall from a height of forty feet will break the bones of any quadruped of that bulk; but it is true that only *overhanging* cliffs are likely to prove fatal to the cimarrón. In descending a steep declivity, or even a perpendicular, but not absolutely straight, rock-wall, he generally contrives to break his fall by taking advantage of every cleft or protuberance large enough to give him a foothold for a moment, and his sharp cloven hoofs seem specially designed for such purposes. Even goats have that trick. I knew a billy-goat that would scramble down a high garden-wall as a bear slides down a tree, and not under the impulse of fear either, but merely to save himself the trouble of a little détour.

The North-Mexican mountaineers hunt bighorns with a special breed of fleet dogs called galgos, or *cimarron-eros*, in Nueva Leon, and said to be descendants of those powerful sleuth-hounds that are used to chase the wolf and the Iberian ibex in the Eastern Pyrenees. In quiet winter nights the cimarróns often descend to the middle region of the sierra, but hurry back to the highlands at the first alarm; and, taking advantage of this habit, the hunting-party divide their forces. A couple of galgos are taken straight to a mountain-meadow where cimarróns are known to graze in the morning; the rest circumvent their retreat and take post at some point of the summit-region where they can watch the movements of the game. At a given signal the first galgos are slipped, and, though they may fail to overtake the fugitives, they will put them to hard shifts before they reach the uplands, where they have to run the gauntlet of the second detachment. If the dogs understand their business, they will co-operate and keep their game together till they can make a simultaneous attack; for, if the herd scatters, the first victim will generally prove a scapegoat for the rest. Going straight up-hill the cimarróns often improve their start by dashing up a cliff where the pursuer has to turn to the left or right, but on level ground the tables are turned, and, once abreast of his game, the hound makes short work of it, dashes ahead of the nearest good-sized sheep,— often a nursing ewe,—and, suddenly turning, flies at her throat in true wolf style and *le rasga la vida*, as the Spaniards express it,—" tears out her life,"—at the first grip. The galgo does not remove his prey, but stays on the spot and summons the hunter by a peculiar howl, repeated at shorter and shorter intervals if he has reason to fear that snow-drifts or prowling wolves will make his post untenable. Professional cimarrón-hunters generally carry a meat-bag, as contact with the hairy coat of the deer-sheep often inflicts the human skin with *cosquillas* ("sheep-tickle"), a persistent itch that sometimes spreads from the hands to the chest, but, strange to say, cannot be traced to any visible cause. Like mange and prurigo, it is probably caused by microscopic parasites.

Dogs can be employed only where the game is very abundant, for, if a band of cimarróns has been chased twice or thrice in the same sierra, they are apt to leave their old haunts forever or become so shy that the pursuit ceases to pay. A herd that has once smelt powder is very hard to get at: their natural timidity becomes a restless distrust, constant practice develops an almost preternatural acuteness of their organs of sight and smell, and they learn to recognize the form of their arch-foe at a great distance and in all possible postures, —standing, crawling, or on horseback. If they can only scent his approach without seeing him or knowing his approximate whereabouts, they instantly decamp to the windward, well knowing that thereby they will either elude their enemy or ascertain his position, preferring to bring matters to a crisis some way or another rather than endure the torture of uncertainty. Among the rocks of a high mountain-region the echo of remote sounds is strangely deceptive; the reverberations seem to come from all sides at once, and on hearing a shot or the boom of a distant rock-blast a whole herd will often resolve itself into a committee of investigation, scattering left and right, scrambling up the cliffs and spying in every direction, then, returning, confer with anxious looks and stamping hoofs, and disperse again till they can agree upon the safest line of retreat. They seem to have some notion of the *modus operandi* of gunpowder, for, if by any chance they meet an armed hunter face to face, they will strain every nerve not only to get out of range in the shortest possible time, but also to confuse his aim by the fitfulness and rapidity of their motion, touching the ground only for a moment, coming down in a wide leap and up again instantly like a rebounding ball, but going zigzag withal: so that the best marksman has to fire at random or content himself with picking off a straggling lamb. A half-hit is as bad as a miss, for an old bighorn takes an amazing deal of killing: a shot through the neck or entrails will not produce any visible effect for the first thirty or forty minutes.

Herons, hawks, and some other birds that cannot hide their nests are sure to select the tallest tree in a thousand, and a similar instinct seems to guide the cimarrón in the choice of his pasture-grounds. He knows what sort of rocks the average hunter would call inaccessible. The North American alps abound with such rocks. Only the roving Apache has ever approached the heights that hide the sources of the Rio Gila. In the Wind River Mountains, in the Wyoming Black Hills, and on the eastern slope of the Sierra Nevada there are thousands of square miles which no hunter's eye but that of Orion has ever surveyed. The town of Monclova, near Monterey, is half surrounded by ramparts of the Sierra de San Simon, and from the bastion of a military post in the neighborhood of the town the soldiers could often see a herd of cimarróns frolicking about and cropping the grass at the brink of an inaccessible plateau. They used to disappear at the approach of the dry season,—on account of the meagre pasture, as it seemed, till it was discovered that in dry summers the plateau could be reached through the ravine of a creek which formed a series of cascades during the larger part of the year. Nearly every herd of our higher sierras has such a place of refuge, which they never approach by a direct way if they can hope to elude the pursuer by leading him a long chase through the rock-labyrinth of the lower cliffs. The ewes of a flying herd invariably bring up the rear, for fear of losing their lambs; and the American sportsman therefore makes it a rule to fire upon the first head in the troop, unless he can single out the males by their broad horns and stouter necks. If this rule were observed by the Mexican hunters it would explain the fact that in the Southern sierras, as well as in the Northwest, old rams are often met alone at a considerable distance from the regular pasture-grounds of their relatives. An old bachelor of this sort is almost unapproachable, and has a knack of disappearing like a mountain-sprite, or manages to frequent the borders of civilization for years before his existence is suspected by the next neighbors. A herd with nursing ewes cannot hide their tracks in that way; what with indiscreet youngsters and anxious mothers, they are too apt to expose themselves at critical moments, and are rarely out of trouble. The old rams seem to know this, and to have come to the conclusion that the safest paths are those which a body can walk alone, and that celibacy is, after all, the best life for a peace-loving cimarrón. Near Granite Gap, Colorado, the surveyors of the San Juan Railroad became familiar with the track of an old bighorn that used to pay a nightly visit to their bivouac in order to share the hay-rations of their ponies; but when they took it into their heads to patrol the camp after dark their guest failed to return, and his spoor was seen no more.

The San Juan range used to be a great hunting-ground for bighorns, and it seems that they are reappearing on the southern slope since the old Utah trail of ante-railroad fame has been abandoned, and that portions of the California Coast Range have thus been repeopled by emigrants from the Sierra Nevada. On the heights of the great central plateau that forms the backbone of our continent the cimarróns will never be entirely exterminated. Their range is too boundless; the extent of the far-western sierras is too immeasurable. Even on a map the maze of winding and intertwisted mountain-ranges, with their net-work of foot-hills, branches, and spurs, is quite bewildering; but only the hunter knows what a sub-labyrinth of highlands and valleys every one of those little shaded streaks represents, what jagged ridges, lateral chains, cross-chains, wide-branching creeks and cañons, plateaux, peaks, and wooded heights, stretching away in every direction farther than his eyesight reaches from the top of the highest rock, measureless alpine systems as intricate in their surface-conformation as the convoluted structure of a walnut-kernel,—all represented on the map by a shaded streak half an inch long and hidden among a net-work of similar streaks.

The incalculable influence of civilization upon the physical geography of cultivated lands makes it difficult to predict the ultimate fate of the wild fauna of a continent like ours; but, judging from present indications, it would seem that the buffalo must perish and that the mountain sheep will survive. The aborigines of the New World were a race of valley-dwellers; among their conquerors, too, the master-nation, the North Saxons, are lowlanders by preference; and in one respect North America will therefore probably remain what our ancestors found it three centuries ago,—a continent of lonely mountain-ranges.

CHAPTER III.

A STEP-CHILD OF NATURE.

THE evergreen hill-forests that cover the border-states of Southern Mexico harbor an amazing number of noisy birds and quadrupeds. All night long the jungles resound with the scream of the tree-panther and the plaintive cry of the *mono espectro*, or ghost-monkey, trumpet-voiced cranes call to each other from the canebrakes, and the deep-mouthed cave-owl booms from the upland thickets. At the first glimmering of dawn the jungle-pheasant sounds his reveille, and long before sunrise the woods burst into a universal chorus of birdvoices, often accompanied by the drumming croak of the tamandua or the flute-signals of the gregarious spider-monkey.

The only pause of the many-voiced concert occurs during the thermal noon, in the first two or three hours after mid-day. In May and June—the dog-days of the northern tropics—even insects need a siesta. When the summer sun reaches the meridian, every animal disappears, and there are minutes when the stillness becomes breathless: the very air seems to stagnate; the leaves droop, as if the pulsations of Nature's heart had stopped.

In such moments the traveller who has sought the shade of the caucho forests is often startled by a singular cry in the tree-tops, a long-drawn, tremulous moan, not unlike the wail of the whippoorwill or a certain lugubrious variation of a watch-dog's yelp. What can it be?—a night-monkey or an owl hooting in broad daylight?

"It's a *tardo*," (black sloth) explains your guide: "he must be somewhere on the south side of that tree. They are very fond of sunshine."

The tardo (*Bradypus tardigradus*) has a peculiar talent for making himself invisible. Even a medium-sized tree, without an excessive supplement of tangle-vines, has to be inspected thoroughly and from different points of view before a slight movement in the upper branches attracts your attention to a fluffy-looking clump, not easy to distinguish from the dark-colored clusters of the feather-mistletoe (*Viscum rubrum*) which frequents the tree-tops of this mountain-region. Closely-resembling clusters of feathery leaves and feathery hair are often seen side by side on the same branch. Which of them is the animated one? A load of buckshot may fail to settle the point. I have seen a troop of idle soldiers bombarding a sloth-tree for half an hour with the heaviest available missiles without being able to force the *strong-hold* of the occupant, who only tightened his grip when a well-aimed stone crushed his head visibly and audibly. But with a good rifle you may dislodge the most tenacious tardo by hitting his branch somewhere below his foothold, for a fractured caucho-stick will snap like a cabbage-stalk. Thus displanted, the falling sloth clutches at the empty air or snaps off twig after twig in his headlong descent, but generally manages to fetch up on one of the stout lower branches, and at once hugs it with all the energy of his prehensile organs; and there he hangs, within easy reach of your arm, perhaps, but without betraying the slightest concern at your approach. The human voice has no terrors for the stoic tardigrade; menacing gestures fail to impress him. A blank cartridge exploded under his nose will hardly make him wink, unless the powder should singe his eyelids. He permits you to lift his claw, but drops it as soon as you withdraw your hand. If you prod him, he breaks forth in a moan that seems to express a lament over the painfulness of earthly affairs in general rather than resentment of your particular act. By and by his love of caloric may lure him back to the sunny side of the tree, but no incentives *a tergo* will accelerate his movements. His claws are a quarter of a foot long and rigidly tenacious, and, once unhooked, he forthwith transfers his attachment to your own person. After spreading his talons fan-shape, he clasps your arm with an intimacy that seems intended to reassure you of his peaceful intentions, but will gradually draw himself well up, as if unwilling to interfere with your locomotive facilities.

Judging from the size of his claws, it would seem that he might use them in a pluckier way; but after a closer examination the sloth can hardly be blamed that discretion should be largely the better part of his valor. His equipment for the struggle of existence evinces, indeed, an almost unfair and certainly unparalleled parsimony on the part of our all-mother Nature. The Bradypus tardigradus has only three toes on each foot and two fingers per hand, making a total of ten claws, to the squirrel's eighteen and the bear's twenty; his legs are so stiff that they can only be laterally extended, and so awkwardly curved that the knees cannot be brought together, thus making his movements on a level surface as hobbling as those of a sprained bat. His molars are very poorly developed, being merely attached to the exterior gums, without roots and without enamel, while the bicuspids, canines, and incisors are entirely wanting. The tail is stumpy or absent, the jaws short, the skull flat and truncated. His eyes are small, and, like his ears, almost buried in tufts of coarse, wiry hair. In short, the sloth is a creature with the vertebrate groundwork of a mammal, but sadly stinted in the "sizings" of nearly all his complementary organs.

The school of Antisthenes, however, demonstrated that a reduction of our wants is virtually equivalent to an enlargement of our means; and, by pursuing this principle to its grim extreme, the tardo contrives to eke out a precarious existence. He is a strict vegetarian, and contents himself with a diet which few of his fellow-creatures are likely to grudge him,—the leathery leaves of the caucho (*Nyssa euphorbia*) and taxus-tree. He sticks to the milky sap of his caucho-leaves, and totally abstains from water and all other seductive drinks. He never indulges in terrestrial rambles, but, like Simon Stylites, passes his life in "aërial penance" on the loftiest tree-tops of the primeval forest, where neither man nor beast can accuse him of trespassing on their domain. The sloth is the only exclusively arboreal mammal. A hill-farmer of the Sierra Madre in the State of Tabasco told me that a family of black tardos inhabited a clump of shade-trees behind his house for eleven years without ever condescending to terra firma or even to the lower regions of their leafy domicile, and often passed weeks and months on the same branch. In the *tierra caliente*, where fig-tamarinds and euphorbias grow to an enormous size, an old sloth may become the hamadryad of a single tree, for, unlike most stupid creatures, the bradypus is a sparing feeder, and, judging from the abstemiousness of domesticated specimens, I should say that four or five ounces of his favorite food represent about the average quantity of his daily ration.

The sloth is as chary of his motions as an orthodox Trappist of his words. Sedate as if he had to give account of every idle movement, he rarely betrays his whereabouts after the manner of squirrels and monkeys, that often become victims to their passion for locomotion. The large cats of the American tropics are not sharp-scented, but hunt by sight in daytime and by hearing at night, and sounds or motions seldom reveal the hiding-place of the discreet tardigrade. In moonlit nights his cry comes from the depths of the virgin woods with a vibratory clang that makes it rather difficult to locate his tree, and even in his honeymoon season the sloth is very taciturn and rarely repeats his call in the same hour. Before sunrise he retreats behind the screen of the liana-shrouds, and remains motionless till the noontide glow has silenced the voices of the forest. On cool days he never stirs at all. He has to give his enemies a wide berth: it is his one chance of safety. By harming nobody and competing with nobody's pursuits, he hopes to enjoy his humble fare in peace.

But, as Stanislaus Augustus said from sad experience, "innocence is no excuse before the tribunal of war," and, in the tropics at least, a state of nature is a state of incessant warfare. In spite, therefore, of all his precautions and his monopoly of an almost unlimited food-supply, the sloth is found nowhere in great numbers; his enemies are too many for a creature that can neither fight nor fly. The harpy-eagle skims the tree-tops of the *tierra caliente* or falls upon him like a flash from the clouds, the lynx lurks in the twilight of the shade-trees, the sneaking ocelot explores the inmost penetralia of the liana-maze: if he meets them, he meets his death. Carnivora have to combine caution with sudden swift-

A NEW DEPARTURE.

ness to catch a monkey in daytime, but sloth - hunting is a search rather than a chase ; small palm - cats or sluggish bears may take a morning ramble through the branches of his chosen tree, and if they espy the poor leaf-eater his capture follows as a matter of course; they need not pursue him, they can collar him at their leisure; a hungry bear collects a family of sloths as he would gather a bunch of grapes.

There is a weasel-like animal allied to the *Mustela*

martes, or pine marten, the *comadrón* (*Martes torquatus*), which haunts the rocks and hollow trees of the South-Mexican sierras and sometimes visits the hen-roosts of the mountain-farmers on its nocturnal excursions. The creature is not much larger than a dormouse, and is dreaded as an egg-sucker rather than as a chicken-thief; but this same tree-rat commits frequent, and generally successful, assaults upon the big tardigrade, and during a visit to Cape Nuna, on the Bay of Campeche, I was shown the skin of a large whity-brown sloth which had been obtained under the following curious circumstances. A party of lumbermen were hauling dye-wood logs from a neighboring swamp, when the barking of their dog and a strange hissing and grunting noise drew their attention to a coppice of rhexia-bushes. On their approach, a pair of comadróns whisked out and bolted up the next tree with a flourish of their bushy tails, but in the underbrush of the coppice and half hidden under a litter of twigs and fresh leaves was found a *tarda morena* with her young, a female sloth of a rare light-brown variety, the youngster dead, the mother *in articulo mortis*. The little one's claws were still clasping the neck of its dam, but its head was nearly gone: the comadróns had eaten its brain and the larger part of its face. The mother's back had been skinned from the rump to the neck, and the hair torn off her shoulders, as if the weasels had tried to get at her throat. When the lumbermen skinned her the hide came off in two pieces, having been gnawed through to the very bone all along the spine. A trail of blood from the coppice to the next caucho-tree told the story of her misfortune. The comadróns had tackled her in the tree-top and worried her till she attempted to escape the best way she could, by letting go and dropping to the ground

PREPAYING THE DEBT OF NATURE.

with the youngster in her arms. But the murderers followed and rode her into the next bush, biting away till they brought her to a stand-still.

Palm-rats and tree-raccoons, too, are apt to try their teeth on the helpless edentate; nay, his near relatives and fellow-vegetarians the marmosets and sapajou monkeys often tease him, or by their indiscreet chattering betray his whereabouts with all the *schadenfreude*— "mischief-joy"—of blabbing school-boys. Even birds join in that heartless sport. The discovery of a sloth seems to excite them like the aspect of a blinking owl. A tardo is as lean as a monkey; the sharpest teeth could not pick more than twelve ounces of meat from his bones; but for the sake of those twelve ounces the South-American variety is unmercifully hunted by the Brazilian plantation-slaves, who have to eke out their meat-rations with tortoise-eggs and such game as they can procure without fire-arms.

No enemy, however, can catch the sloth napping; his is a sleepless soul; his inert brain requires no rest. Heat and cold do not affect his sensorium; you may see him hang on to a top branch under the glare of a vertical sun, eating placidly,—listless and mute, like a survivor of the antediluvian fauna, the age of sluggish monsters, when Professor Owen's sloth-like megatheriums pastured the fern-forests of the tertiary period.

But if his physical organization classes the sloth with the lowest mammals, his mental calibre degrades him below the rank of a first-class reptile. There is a small Peruvian variety of arboreal tardigrades, the *unau*, or spotted sloth, whose habits in captivity I had no opportunity to observe; but in the brain of the *tardo real*, the large dark-brown sloth of Mexico and Central America, the faculties which distinguish the average mammal from a mollusk are either undeveloped or wholly extinct. The proprietor of the Hotel de Cuatro Naciones in Puebla owns a three-legged sloth which he domesticated

A SLOTHFUL FAMILY.

in a little kitchen-garden six years ago ; and, though fed daily by the same hands, the old pensioner still fails to identify his benefactor or to recognize his obligations in any way. To his ear the human voice in its most endearing tones is a grunt *et præterea nihil:* you might as well appeal to the affections of a cockroach. You may frighten a pig, a goose, a frog, and even a fly, but you cannot frighten or surprise a sloth. On my last trip to Vera Cruz I procured a pair of black tardos, full-grown and in a normal state of health, so far as I could judge, but after a series of careful experiments I have to conclude that their instinct of self-preservation cannot be acted upon through the medium of their optic or acoustic nerves. They can distinguish their favorite food at a distance of ten or twelve yards, and the female is not deaf, for she answers the call of her mate from an adjoining room; but the approach of a ferocious-looking dog leaves her as calm as the sudden descent of a meat-axe within an inch of her nose. The he-sloth witnessed the accidental conflagration of his straw couch with the coolness of a veteran fireman. War-whoops do not affect his composure. I tried him with French-horn-blasts and detonating powder, but he would not budge. One of my visitors exploded some pyrotechnic mixtures of wondrous colors and odors, but the tardo declined to marvel: he is a *nil-admirari* philosopher of an ultra-Horatian school.

He has learned to accelerate his progress on a level surface by sliding on his haunches, using the claws of his forefeet like grappling-hooks, and thus crawled one day into a basket that had been assigned to a nursing fox-squirrel and her infant family. She flew at him like a little bull-dog, gave him a snap-bite, and then stood at bay, chattering and switching her tail, but repeated her assault whenever he stirred or as much as turned his eyes in the direction of the nest. The tardo grunted a feeble protest, but offered no resistance, and finally seemed to accept this new phase of his existence as a dispensation of inexorable Fate. The idea of evacuating the basket never suggested itself to his guileless soul.

My exotic guests have taken their summer-quarters in an old tool-shed with a more or less happy family of indigenous pets, squirrels, gophers, and black-snakes, and the conduct of the smaller boarders at first evinced their deference to the superior size of the foreigners; but they soon learned to ignore their very existence, or to treat them as locomotive vegetables, whose rights no superior being need respect. The gophers use them as jumping-boards, and usurp their couch with a cool disregard of preëmption-laws; the black-snakes sun themselves on the broad back of the he-sloth, and one of the squirrels has no hesitation in providing herself with nest-building material from his hirsute hide. I have seen a gopher pluck bits of half-chewed apple-peels from the jaws of the patient tardos; and I believe they would submit to excoriation if one of their neighbors should be in need of a fur cap.

There is no fun in a sloth; his motions are limited to a few indispensable functions, and he performs them like an ill-constructed automaton. He does not appreciate caresses; practical jokes delight him not. Even the young ones have nothing of the vivacity and playfulness of other infant mammalia. Grotesque little imps, with woolly heads and preposterous claws, they will cling for hours to the rump of their parent, with their noses buried in her fur, never vouchsafing the external world a look or sniff. Yet they can be easily weaned, and will cling as tenderly to a "sham mother,"—a milk-bottle enveloped in a piece of fluffy cloth,—their attachment to their natural nurse being merely that of a suctorial parasite to its victim. They develop very rapidly, in weight, at least, for in agility and intelligence the new-born *tardillos* are faithful copies of their full-grown progenitors. Their private life as well as their functions in the household of Nature could be successfully enacted by a big caterpillar. I have often watched my tardos when they thought themselves unobserved, and I do not think that the conduct of a starfish could be more exclusively controlled by what biologists call the "blind instincts." They will ensconce themselves in a corner or squat down in the very centre of the shed, as chance directs, and there they sit, not asleep, but contentedly inert, in the languor of idiocy, for hours and hours. If their door is left open on a chilly morning, they sometimes come out to enjoy the sunshine at the rear of the shed, but, instead of taking a bee-line toward the door, they will crawl along the walls and nose around in the corners in a manner strangely suggestive of the movements of an imprisoned beetle. They have a curious fashion of making their way to the very top of every ascendible object, the back of a chair or the elbow of a stove-pipe, and out in the garden often pass the larger part of the day on the knob of a gate-post, brooding perhaps over dreamy mementos of their lost tree-top paradise. If their prison is closed, I have seen them raise themselves on their hind-legs and inspect a piece of clothes-line depending from a nail near the door. They often cast wistful glances in the direction of that rope,—why, I know not, since they are too clumsy to climb it; but I suspect that they would like to get away and go home. The rope possibly reminds them of the bush-ropes dangling from the canopy of their native cauchos. Nostalgia, or rather a vague yearning for freedom, may be the one touch of Nature that makes even the sloth akin to the rest of mammal kind.

In spite of their formidable claws, they are by no means first-rate climbers: they can hook their way along a horizontal bar and scale a ladder or an arm-chair, but are unable to climb a smooth rope or a smooth slender tree. I once put them on the crook of a young apple-tree, to see if they could make their way to the upper branches, but, after clawing away at the stem as if trying to find some notch or protuberance, the male came down head foremost, and vented his shocked feelings in a rasping grunt.

This grunt and a feeble, parrying movement of his fore-legs seem to be his only means of self-defence,— a *dernier ressort*, reserved for emergencies. If a dog bites him, or if you offer him a tidbit after a prolonged fast and snatch it away from his very jaws, he will slowly turn his head, and then, as if the significance of the indignity were gradually dawning upon his mind, he breaks forth into crescendo grunts, resembling at once the whirr of a buzz-saw and the droning hum of a bee-hive. I do not know if a sloth can be teased into active resistance, for, after trying all my conscience and Mr. Bergh would permit, that point still remains undecided. A Spanish-American sportsman, however, told me that the females sometimes use their claws in defence of their young. This would seem to prove that not resentment or even self-preservation, but child-love and the love of freedom are either the most radical or the most inalienable instincts of the animal mind. The vivacity of an animal does not depend exclusively on the perfection of its motory organs, for there are sluggish birds and restless reptiles, and the sloth, too, is lazier than even his clumsy structure seems to warrant. His fur is infested with various parasites, but he never employs his long claws in entomological pursuits. On principle rather than from absolute helplessness he appears to surrender at discretion to all his enemies, great or small. I do believe that a swarm of horse-ants could eat him alive without meeting with any serious objection on his part. He holds his own life cheaper than that of a sand-flea.

It looks, indeed, as if neither the sustaining nor the creative agencies of Nature had thought it quite worth while to exert themselves for the benefit of the poor tardo; Vishnu may have deemed it a waste of trouble to devise safeguards for the preservation of a life of so little value even to its possessor. For what should endear existence to a creature that passes its days in purblind apathy, in a vegetable torpor, incapable alike of mental and physical activity? The instincts of a sloth are those of a cuttle-fish; the sense of frolic and the sense of comfort are not represented by any organ of his cranium. He never sleeps, but his vigils are not those of a wide-awake creature: his life is a long trance of open-eyed inanity. Even "alimentiveness," the sole solace of many brainless beings, seems to him but a scanty source of enjoyment. His process of mastication is slow and laborious; he cannot gorge himself with his toothless jaws.

Still, Fate has granted the much-bereft edentate one compensation,—a cheap one, indeed, but still an offset to many defects: a most contented disposition. On the morning of an unusually cold April day I was summoned to a neighboring town, and took a look at my tool-house menagerie before I left. Finding that the female sloth had monopolized the family couch, I carried her mate up to an empty garret and attached his claws to a mantel-piece where he could warm himself by putting his back against a flue of a hot-air chamber. An unexpected delay prevented my return that night, and when I got home the next morning I entered the garret with sore misgivings about the survival of my tardo. But no; there he hung, on the very same spot and in the same attitude, imbibing caloric at every pore, and purring to himself in dreamy beatitude,—a tardo temporarily satisfied that life was worth living.

Like poor Lo, the sloth has no friend to rely on and but little talent for self-help, but if his desires are limited to sunshine and caucho-leaves he need not complain. Our well-being, for all we know, may depend less on the nature of our wants than on their proportion to our means, and the bug whose necessities can be supplied by crawling from leaf to leaf is possibly as content as the bird that wings its flight from tree to tree.

Yet this negative kind of happiness seems somehow incongruous in a creature so nearly allied to the primates of the animal kingdom, so that even from this point of view the sloth may be considered as an abnormal phenomenon,—a combination of a vertebrate form with the mind of an insect.

CHAPTER IV.

SECRETIVENESS.

ANIMALS in a state of nature are endowed with certain protective instincts to a degree which might often tempt us to believe in the existence of a "sixth sense:" the clairvoyance of bats, for instance, and the topographical second-sight of migratory birds seem almost too marvellous for any less mystic theory. But the specific purpose of such instincts, and their great variety in degree as well as in kind, make it more probable that in stress of circumstances any one of the more or less rudimentary faculties which the lowest animals share with the highest is capable of an almost infinite development. The necessity of pursuing its prey under water has taught the dap-chick to find insects at the bottom of a muddy creek. The exigencies which compel a nursing she-wolf to return by the shortest route from a hunting-expedition in a distant mountain-range have perhaps endowed the ancestors of the genus *Canis* with that marvellous faculty of direction; and the defensive warfare of many animals against an enemy of superior strength may have developed their instinct of caution to the degree which enables them to hold their own even against the *Simia destructor*,—the terrible weapon-inventor, with his murderous machines and four-footed allies.

Phrenologists assure us that the skull-bones of all mammals indicate their mental characteristics; and I have often examined the head of a weasel and wondered at the flat top of its little occiput, for, if Spurzheim is right, the "organ of secretiveness" should form a protuberance resembling the horn of a Texas toad. A caged weasel seems rather an uninteresting pet, slow at learning tricks and not very quick in distinguishing playthings from comestibles; but let it escape in a furnished room, and its peculiar *forte* will marvellously assert itself: it will vanish at once, and, with the curious felicity of some lawyers in hitting impromptu upon the one tenable subterfuge, it will at once get into the very best hiding-place the territory affords,—the lining of an old dressing-gown or the interior mechanism of a spring-mattress,—and remain invisible and almost inaudible for days together. Squirrels, too, have a curious knack of disappearing at the critical moment and keeping out of sight, even on leafless trees, by dodging behind branches and excrescences that seem hardly large enough to hide a good-sized mouse. Our little Northern gray squirrel rummages the penetralia of every hollow tree, but for its family nest it almost invariably chooses a cavity opening into what lumbermen call a "fork-split,"—*i.e.*, a crack in the fork or saddle between the stem of a tree and one of the main branches, and, besides being invisible from below, such holes are generally stopped with moss and leaves. North of the Arkansas the black bear passes the three coldest winter months underground, either in a cave or in a "dug-out," generally in the deep vegetable mould near the roots of a fallen tree, and it is only by the sheerest accident that such burrows are ever discovered, though the old hibernator leaves a very visible spoor and is not over-particular about covering the rear of his shaggy fur. He relies on his talent for choosing the site of his dormitory, and is sure to select the most unfrequented spot in a wide labyrinth of valleys and mountain-ranges. The lower glens, with their sheltered coves and perennial rills, must be very tempting; but black cattle and their proprietors are apt to visit such places on cold winter days, while hunters find the best trails along the ridges, on the very backbone of a mountain-chain. Ursus niger therefore prefers the middle region, some wild ravine in the steepest rocks of the mountain-flank, half-way between the ridge and the sheltered valleys, and, if possible, on the eastern slope, because this side of the Sierra Nevada the coldest winds come from the northwest.

But the unrivalled masters in the art of nest-hiding are the feathered songsters of the sparrow tribe, the *Passeres*, as our ornithologists call them, though in this respect the sparrow himself can hardly pass for a representative bird. His cousins, though, the linnets, finches, and ortolans, cannot reproach themselves with risking the loss of their nestlings by neglecting any possible— certainly not any humanly possible—precautions in the construction of their little nurseries. The nest of a greenfinch on a willow- or chestnut-tree defies detection, unless you should happen to espy the bird in the act of feeding her young. No botanist could more exactly match the color of the tree-bark with the blended hues of lichens and dry bind-weeds, or imitate with interwoven twigs the characteristic forms of the protuberances and "knots" of a gnarled branch. Viewed from below, the grass bower of the Italian ortolan cannot be distinguished from the grayish-green tint of a half-withered olive-leaf; the bag-nest of the golden wren is hidden amidst the drooping tassels of the mountain-larch; and the pendulous cradle of the orchard oriole looks exactly like the accidental excrescence of an old apple-tree.

Birds that cannot imitate such textile masterpieces show a consummate skill in foiling their enemies,— hawks, cats, and boys. Our tanager never takes a bee-line to her nest, but beats about the bush, apparently in search of food, till she sees an opportunity to slip in unobserved. It is well known that quails and many small birds often try to divert the attention of a nest-robber by throwing themselves in his way, shrieking and imitating the movements of an unfledged nestling; but it is a curious fact that in such critical moments the nestlings themselves keep perfectly quiet, for hours, if necessary. I once examined the nest of a North-Carolina chatterer (*Ampelis garrula*), and while the

A VANTAGE-GROUND.

parent couple fluttered to and fro with incessant screams their young ones kept as still as mice, though generally the approach of their mother was greeted by a chorus of voices that justified the surname of the species.

Squirrels, pheasants, and some other cautious nest-builders show the same ingenuity in choosing an impromptu hiding-place; wounded partridges often crouch motionless between the twigs of a small bush where no *a priori* philosopher would suspect their presence. Colonel S——, a great hunter before the Lord, and proprietor of a sylvan Tusculum near Huntsville, Alabama, in repairing the lath-work of his vine-arbor happened to inspect a part of the roof that had never been visited by the grape-gatherers, when down jumped an animal which the astonished Nimrod recognized as a black-tail fox, an old offender, for whose special benefit the sporting fraternity of the county had been under arms that very morning. No one had dreamed of beating the enclosed garden, and Master Black-tail had evidently ascertained by experience that *procul de Jove procul de fulmine* is a rule with occasional exceptions. Musk-rats, too, have a knack of burrowing in the most unexpected spots of a frequented river-bank. Indeed, all much-hunted animals seem to find by a sort of intuition the safest and *most out-of-the-way* places of refuge. This instinct may, after all, furnish the right solution of a problem that has puzzled more than one speculative philosopher,—the question, namely, where animals bury their dead:

> What becomes of pins, we should like to know,
> And the birds that die, where do they go?

What becomes of old birds? Do they perish in the attempt to cross the sea in their biennial migrations, or are they eaten, devoured utterly by beasts of prey or insects? Why is it that hunters so rarely come across the remnants of a murdered bird? the ravages of hawks and owls do not account for the tenth part of the mortality implied by the difference between the possible and actual yearly increase of creatures that rear from five to ten young ones every spring. And what about the larger habitants of the wilderness, the countless Polish wolves, Georgia raccoons, and Texas squirrels, that do not fall by the hand of man and have few other enemies? Are they eaten by ants? The woods would be covered with skeletons. No; I believe that animals die in the best hiding-places they can find, and in a region of tangle-woods and craggy cliffs that would mean a good deal. Some of the bones we find in the Jurassic limestone caves are perhaps the remnants of animals whose secretiveness *in articulo mortis* was rewarded by an undisturbed repose of fifteen or twenty thousand years. Raccoons, like bears and other plantigrades, are subject to a kind of mange, and one of my acquaintances in Columbus, Georgia, tried to cure his pet coon with a dose of nux vomica. The remedy promised success if life itself is a disease : the little plantigrade stretched his legs and his eyes turned yellow. After sunset, however, on being exposed to the cool night-wind, he revived and crawled away,—to the sanitarium of the wilderness, it was supposed; but two years after, his remains, known by a rusty wire collar, were discovered below the effluent pipe of a hot-house, which he could have reached only by digging under the foundation-timbers.

The strategists trained in the school of the Corsican campeador cannot have chosen the moment of attack with a closer calculation of all circumstances than the nocturnal prowlers that prey upon our orchards and hen-roosts. Their practice seems to contradict the idea that people sleep soundest before midnight: between one and two A.M. is the ghost-hour,—the time when most Christians are actually asleep, and foxes, weasels, and burglars most wide-awake. During a vacation of my boarding-school years I once helped to watch a poultry-house whose tenants had begun to disappear in a way that suggested the agency of a domestic traitor. From ten till twelve the barking of numerous curs facilitated our vigils; soon after midnight the late guest of the village tavern extinguished his candle; but still there was in the air that something not traceable to any single cause and resulting rather from the co-operation of many small sounds,—the creaking of a gate, the distant rumbling of a night-coach, a whispered dialogue. But at one o'clock all was still, and half an hour later all but the low snoring of my companion, when a moving shadow attracted my attention to the top of a high board fence, and there and then came

RECONNOITRING.

the thieves,— two mountain-brook minks, that approached the hen-house with noiseless steps and entered by an unsuspected aperture between the top of the board wall and a loose shingle,—the only loose shingle on the roof, as was afterward ascertained by a thorough examination of the building. The father of the Wagner-worshipping monarch of Bavaria often diverted himself by suddenly lighting the calcium-lamps of the Hof-park after a public *soirée champêtre;* and if some nocturnal landscapes of North America could be illuminated in the same manner, we should find the woods and fields swarming with animals where not a living thing is to be seen in the daytime.

Migratory birds, with few exceptions, travel at night: like terrestrial tourists, they have their favorite routes, their St. Bernard passes and trans-continental highways, and in clear November nights the farmers of Paso del Norte often hear the trumpet-notes of a large flock of wild swans, marshalling their host at an inconsiderable height, to judge from their loud rallying-signals at sight of a conspicuous landmark. Ducks and divers, too, do most of their flitting after dark, and only the champion flyer of all aquatic birds, the wild goose, ventures to travel in the daytime, at an elevation where she can defy the artilleristic machines of her arch-foe.

There is no doubt that wild birds learn to keep the run of the weekdays and leave their cover only at the sound of the American church-bells, while the ringing of the French and Spanish campaniles sounds a death-knell to the feathered inhabitants of non-sabbatarian Europe. The partridges of our Southern States seem to understand even the meaning of a dinner-bell, since the proprietor of a south-Virginia strawberry plantation told me that their depredations absolutely nonplussed him, till he ascertained that they entered his field during the noontide hour, while the gardeners were taking their siesta.

SECRETIVENESS. 107

Not a grief-deriding goddess on a pedestal but a rat at the entrance of his hole would be the fittest emblem of patience. Near the river-warehouses of St.

SUNDAY MORNING.

Louis, Missouri, and on the docks of Galveston Island, you can see their patriarchs mounting picket-guard between sunset and twilight, —gray old sharpers, not easily to be distinguished from the surface of the dusty wharves, and still less by any voluntary motion, unless you know their haunts and keep your eye on their holes. There they sit, tail and hind-legs tucked out of sight, their fore-feet close together, every muscle braced for an immediate sally, yet rigidly motionless as long as there are any boys or dogs in sight. But watch them from behind a closed window or from a perch on the cotton-bales: the moment the road is free, the end of the sharp nose begins to work, the hind-legs become visible, and with a sudden rush, in a perfect bee-line, the rat is across the street and into his subterranean hunting-grounds, perhaps the vault of a bonded warehouse that has lately received a consignment of New-Orleans molasses.

Rats do not like to cross an open street, and mice avoid the centre of the floor if they can possibly reach their objective points by running along the walls: instinct seems to tell them that alongside of a vertical surface the visibility of a body cannot be aggravated by its shadow. They also seem to know that monotonously repeated sounds are least liable to attract attention: the steady gnawing of a wall-mouse "blends with silence" as readily as the ticking of a clock. Co-operative mice keep time: at the end of an interesting chapter I have often become conscious of the fact that a rodent committee of ways and means were rasping away vigorously in my immediate neighborhood and had evidently been hard at work for some time. I have two such partners in my bedroom, and in sleepless nights I sometimes enjoy the overture of their duet. It begins with a pianissimo nibble, a tentative prelude that can be nipped in the bud by projecting a bootjack against the wall, but once fairly started they keep at it and scrape away with a wonderful uniformity of intonation and accent. In the Southern coast-States, where white pine forms the staple building material, rats sometimes undermine a house completely before their presence is as much as suspected; and the destruction by the late tornado of a sea-side boarding-house near Brunswick, Georgia, disclosed a colony of *mephites Americanæ*,—North American skunks, in fact, —that had lined their basement-quarters with a quantity of moss and cotton which must have taken them at least a year to accumulate, during which time they had strictly refrained from all impolite acts. On being confronted with their irate landlord, however, they resented his aggressive conduct in their characteristic way; and if such waters as the Mediterranean and the Gulf of Mexico could be suddenly drained, I believe that the visions of the Ancient Mariner would be realized.

In well-wooded countries even larger animals can survive and thrive for years without betraying their existence,—least of all to their next neighbors, for the fox is not the only robber who spares the vicinity of his headquarters. Only in three of our thirty-eight States— Indiana, Rhode Island, and Connecticut—have bears been entirely exterminated; in all the rest they are either autochthones of the soil or pay occasional Christmas visits, which, like those of other eupeptic relatives, are apt to extend to the end of the winter. Panthers are still found in twenty-six or twenty-seven States, the doubtful twenty-seventh being South Carolina, where some of them are supposed to lurk in the highland gorges of Pickens County. In Texas and the Northwestern Territories wolves are still too plentiful to risk a high scalp-bounty; east of the Mississippi they occur only sporadically, in North Carolina chiefly, and in the wild border-counties between Tennessee and West Virginia. But who would suppose that in the *soi-disant* birth-land of civilization—in France—several hundred of them are killed every year, and not in the Pyrenees merely, but on the Belgian frontier and in the Western Cevennes, hardly two hundred and fifty miles from Paris!

Even in Western Germany the wild fauna of the mountain-regions is by no means confined to the relatives of the sleek preserve-pets. About fourteen years ago I paid a visit to the famous Salzbad of Allendorf, near Cassel, where two of my former school-mates were surveying a tramway from the salt-works to the coal-pits of the Kaufunger-Wald. The surveyor and his assistant were both Hainault men, born and bred in the wilds of the Ardennes, and on being called to a place in the woods where a spoor in the fresh snow had puzzled all their workmen, they at once recognized the track of an old lynx,—a rather frequent visitor to the sheep-folds of the South Belgian mountaineers. The snow was the first of the season; but a few days after, the same spoor was found in a drift at the foot of a ravine, and a sharp-eyed lad traced it to a gypsum-cave, the Salpeter Loch, so called from the nitrous deposits in one of its ramifications. The news soon spread to the Salzbad and caused an animated controversy among the sportsmen of the neighborhood. But in Kurhessen every township has its Oberförster, the overseer of the government forest and the supreme authority on all questions pertaining to woodcraft and venery; and the Förster of Allendorf ridiculed the lynx-report. "There are no lynxes in the Kaufunger-Wald," was his verdict, "except near Almerode, twenty miles from here, and there only in very hard winters." Besides, lynxes and cats stick to the trees till after Christmas, when the mountain-brooks freeze for good; and the Förster ought to know, being a graduate of the Austrian *Forst-Schule* of Hermanstadt in Transylvania, where lynxes are as common as squirrels.

But the exponent of the Ardennes party was equally positive, and on the first sunny day (also the first day of the week, I am sorry to say) we all went to the Salpeter Loch to settle the dispute of the two zoological dogmatists. The ravine being on a government preserve, nobody was permitted to carry arms but the Oberförster, who had shouldered his shot-gun in deference to regulations, though without the least idea of having to use it that day. At the entrance of the Loch the boys had piled up a large heap of brushwood, and upon our arrival the pile was lighted,—merely to please the strangers. The presumptive tenant of the cave might be a badger or a *Feldkatze*,—*i.e.*, a domestic tom-cat run wild,—said the Förster, who had taken a seat on a treestump, but certainly not a lynx, whose short legs oblige him to secure his prey by a downward spring, and whose favorite haunts, therefore, are leafy trees overhanging a spring or a salt-lick.

But, while the professor lectured, the flames rose higher and higher, and in the midst of a dissertation on the habits of the Transylvanian lynx the lecturer was interrupted by a fat specimen of the indigenous variety bouncing from the cave and away through the bush,—taking him so completely by surprise that he stared after the phenomenon in mute bewilderment, and even kept his seat on the tree-stump. When the fugitive had got a start of some eighty or ninety yards, the Förster stood up and fired both barrels after it, two well-aimed shots, for the lynx broke down, rather against my expectation, though the marksman only wondered that it got up again and continued its flight. But this was to be a day of surprises: the smoke of the second shot had not yet cleared off when two more lynxes bounced from the cave, and, rushing through the underbrush, followed their predecessor with superfluous haste.

"You give in?" inquired the surveyor.

"Yes; I have to," said the Oberförster; "but"—after a pause—"from this day on I won't believe in impossibilities any more. No, sir, not I. Before we get through with this survey I should not be the least bit surprised to discover *a troop of secret camelopards!*"

CHAPTER V.

BATS.

DURING a foot-tour through the Western Jura I once saw a crowd of people in a cutting beneath a railway-bridge, and, clambering down the embankment, found that the workmen had exhumed the fossil remains of a gigantic pterodactyl,—a monster with the head of a crocodile and the wings and claws of a bird. As bone after bone was picked out of the gravelly detritus, one of the engineers arranged the skeleton in anatomical order; and I still remember the expression of a peculiar speechless interest on the faces of the spectators: even the Savoyard navvies stood around mute, thrilled with the spell of a by-gone wonder-world.

A similar feeling has often come over me at the sight of a captive bat, wrapped in the folds of its leathery wings or wriggling on the floor in uncouth contortions, and still more vividly in the twilight of an ice-bound cave where I once saw a mass of winged dormice hanging together in a clump, motionless, and answering my voice only by a feeble squeak, like the Lemures in Hesiod's Tartarus. A bat is a living anachronism; there is something obsolete and paradoxical in every part of its organization. Skin wings were quite in vogue in the days of the Devonian monster-period, but have gone out of fashion among the representative creatures of our latter-day world; and it is a curious fact that all winged mammals have become nocturnal, as if they could not compete with the talents of their daylight contemporaries. The winged lemur (*Galeopithecus volans*), the flying fox, and the flying squirrel are all moonshiners, and dread sunlight as miracle-mongers dread the light of science; but they all have the exaggerated optics of an owl, evening-eyes, that catch every ray of the fading twilight, while the eyes of the bat proper are as rudimentary as those of a mole or of the strange fishes that were discharged from the subterranean tarns of Mount Cotopaxi.

Its *sensitiveness*, on the other hand, is developed to a degree that far transcends the functions of what we generally call the sense of touch. Spallanzani demonstrated that blinded bats can fly around a room for hours without ever touching the walls or ceiling; but the faculty of guessing, without actual contact, the proximity of a solid obstacle is shared by other animals: Canadian night-hunters often hear a moose going at top-speed through a thick forest; and a blind horse will stop within a few inches of a barred gate. A greater riddle, however, is the question how bats find their food. Is it possible to imagine that they *feel* the approach of a little beetle meeting them in their rapid flight?—for that they do not hunt at random, like a whale charging open-mouthed into a shoal of herrings, is proved by their quick turns and dodges in pursuit of an individual insect. Nor can their pygmy eyes help them much. In seizing their prey, the jaws of a bat produce a peculiar clicking sound; and I have heard that same click at midnight in the deepest gloom of a tropical forest. The long-eared varieties may *hear* many things that would escape a human ear, but their capacity of finding so much food in the dark is still almost incomprehensible, for most bats are enormous feeders: the Kalong eats twice, and the common horse-shoe bat at least four times, its own weight in the course of the twenty-four hours, and they all have that strange musky odor that seems a characteristic of so many voracious creatures,—the ichneumon, the racing beetle, and the alligator.

As the Euclidean *punctum* is defined as a point without extension, the voice of a bat might be called a sound without vibrations,—a shrill, sudden squeak, unlike any other sound in nature or art. Though piercing enough to be heard from afar, it is too abrupt to guide the ear in any special direction: you can put a wood-bat in a narrow box, and the box on the table, and bet large odds that the incessant shrieks of the captive will not betray its hiding-place: to nine persons out of ten the sound will seem to come from all parts of the room at once.

Many of their habits, too, distinguish the cheiropters from all other creatures of our planet. A r i s- t o t l e classed them with the birds; and in one respect they might even be considered the re presentatives of the class, being, *par excellence*, creatures of the air. All winged insects can run or hop; the sea-gull runs, swims, and dives; but, with the sole exception of the Javanese Roussette, bats are completely "at sea" in the water and almost helpless on terra firma; they eat, drink, and court

CHILDREN OF EREBUS.

their mates on the wing, and the *Nycteris Thebaïca* even carries her young on her nightly excursions. Nay, bats may be said to sleep in the air, for they build neither day-nests nor winter-quarters, but hang by the thumb-nail,—touching their support only with the point of a sharp hook. But this hand-hook connects with muscles of amazing tenacity: in cold climates, where bats have to club together for mutual warmth, fifty or sixty of them have been found in one bundle, representing an aggregate weight of about fifteen pounds, all supported by one thumb-nail. The "head-centres" must sleep as warm as a child in a feather bed; but it is hard to understand how the outsiders can survive the cold season, for, in spite of its voracity, the bat accumulates no fat, and the flying-membrane is a poor protection against a North-American winter. The only explanation is that their winter torpor is a trance, a protracted catalepsy, rather than a sleep: hibernating bears and dormice get wide awake at a minute's notice, but I have handled bats that might have been skinned without betraying a sign of life and needed more than the warmth of my hands to revive them, for their wings were quite brittle with rigid frost. Bats prefer a cave with tortuous ramifications that shelter them against direct draughts, but still with a wide, though not too visible, opening, as they do not like to squeeze themselves through narrow clefts. A dormitory combining these requisites is sure to attract lodgers from far and near: the northern entrance of the tunnel-grotto of Posilippo and the Biels-Höhle in the Hartz are tenanted by hundreds of thousands of bats that avoid all the neighboring caverns; and our Mammoth Cave, with its countless grottos, has only two bat-holes, whose occupants have never been known to change their quarters.

Canadian bats hibernate from six to seven months, without food or drink, and without changing their position by a single inch; but a trance-sleep may come natural to a creature of such limited brains; as a French lady said of a dying *borgne borné*, "he hasn't got many eyes to close, *et point d'esprit à rendre.*" Phrenologically, the cheiropters stand at the bottom of the scale: the frontal bone of a hog is perfectly flat, but that of a bat is *dished*,—bulged the wrong way: its facial angle can be measured only by negative degrees. It would be about as easy to brain a fly as a bat; but, like flies, cheiropters can boast of a remarkable presence of what mind they are gifted with: it is really impossible to hit a flying bat with a stick; in a closed room he will baffle the tactics of a whole broom-brigade for minutes together: the word *subterfuge* must have been derived from his marvellous knack of dodging a blow by a sudden sideward and downward swoop. It has been said that the art of flying will ultimately be learned from bats instead of birds; but I believe that an artificial wing would bear a closer resemblance to a callow feather apparatus than to the sensitive membrane whose net-work of nerves may possibly be the supposed sixth sense of the artful dodger.

In summer the cheiropters of the temperate zone pass the day in hollow trees, under the eaves of old roofs, and even in the interior of open buildings; the landlord of the Salzburg Acropolis has a large "bat-rookery," not in the old burg, but in the loft of an adjoining frame house, whose basement is used for a tenpin-alley, while the loft itself is occasionally smoked out, to treat visitors to an Acherontic spectacle,—a surging cloud of flopping and squeaking imps of darkness.

Bats can be domesticated, but never tamed; in daytime, especially, their sharp teeth are always apt to fasten in the hand that feeds them. Children of Chaos, they love darkness and solitude, and their independence is a practical satire on the arrogance of the self-styled autocrat of the animal kingdom: their whole appearance proclaims the *alter ens*,—creatures that have no part with us and ours. The natt-backa—"night-bird" —has never been a favorite of folk-lore: the myth of the Edda makes it a messenger of Hel, the goddess of darkness and death; and in Oldenburg its sudden appearance in daytime is still considered a fatal omen:

> Nat-bör am Morgen
> Bringt Unglück und Sorgen;

and the Frisian *flederdyn* (Yorkshire "flittermouse") is a synonyme for a wraith or a night-hag. The bat-

epithets of the Eastern nations are equally opprobrious, though the Arabian *gessim-al-sheytan* ("devil-birds") refers exclusively to the ugly Megaderms, or bull-dog bats. The Chinese admire their own death-head profiles, and compare the European nose to the beak of a vulture; Captain Baldwin even mentions a tribe of Zambesi Caffres who deem it unbecoming to wear front teeth, and a he-bat may think his mate a winged Venus; but in the eyes of a Caucasian, at least, the face of a Megaderm seems a combination and aggravation of everything we call hideous,—a wide-split mouth, whose bull-dog lips still fail to cover the greedy teeth; a pug-nose, so *retroussé* that its upward bent forms a twisted hook; pig-eyes, with wrinkled lids; and ears that exaggerate the jackass-pattern by being joined in the middle, thus forming a sort of hood or scalp-flap. Compared with such features, a frog's head appears quite human, a monkey-face almost classic.

The Low-German *Speckmaus* expresses a wide-spread superstition. "Bat, bat, fly in my hat, bring me some bacon-fat," sing our children; and Good-man Hodge will have it that the flittermouse visits his chimney in quest of smoke-meat. But the Spanish farmer adds a more serious charge: besides stealing bacon, the *murcielago* is a cockatrice, and for fear of her evil eye children sleeping in the open air have to cover their faces. Our Mexican neighbors kill all cheiropters with indiscriminate zeal; but farther south that aversion is almost justified; the insectivorous achievements of other bats cannot atone for the sins of the vampire. To many people the musical preludes of a mosquito seem to aggravate the hatefulness of its visits; but the absolute noiselessness of a vampire is a great deal worse: a tickling sensation, becoming gradually stinging and painful, or the dripping of a blood-drenched hammock, is the first indication of its presence, and to persons of a nervous temperament the mere suspicion of that presence is almost intolerable. Near the haunts of the ghoul-bat a flitting shadow on a moonlit wall is often sufficient to banish sleep for the rest of the night. In the lowlands of the tropics the airiest bedrooms are generally the most popular, and where people sleep in the open air the vampire has it much his own way. Veils and gauze stockings, however, afford at least a partial protection, by obliging the blood-sucker to use his teeth instead of his tongue, and thus awakening the sleeper in time, the painfulness of the preventive being outweighed by the pleasure of revenge,—"un piacer che vaglia mil tormentos." I knew an apiarist who carried business, or Buddhism, to the length of "easing off" a stinging bee instead of smashing it; but Uncle Toby himself would not have spared a captured vampire.

Bonpland recommends an ointment of peppermint oil, and the Guahiba Indians of the Lower Orinoco post a sentry,—a watch-dog who has to pass the night in a basket suspended from the lintel of the open door. To a sleeping dog the winged incubus probably betrays itself by its teeth rather than by its odor; though there is no

A VAMPIRE-TRAP.

doubt that even the Indians can *smell* the approach of a vampire, and the negro servant of my travelling companion in Central America often horrified us by manifestations of the same faculty. "*Ben attention!—Je sens un chorussi*" (corrupted from *chauve-souris*), he would bawl out in the middle of the night, and the flash of a nitre-match rarely failed to justify the warning by the testimony of our eyes.

There are four or five species of vampires in the American tropics. Azara holds that none of the indigenous animals are plagued by these pests, nor by mosquitoes either. But thereby hangs an enigma: granting that the fur of a bear and the feather mantle of a bird are impervious to the sting of a tipulary insect, what do they all live on, the countless gnats that never get a chance to commit phlebotomy? In the "Sunken Lands" between Memphis and Little Rock it would be a moderate estimate to say that there must be a million mosquitoes to the square mile. What do they all do for a living? Do they live on hope and one bite a year, or are they vegetarians whose appetite, like ours, is subject to sanguinary aberrations? So much is certain, that the vampire has all the physiological characteristics of an insectivorous bat, and if his blood-thirst should be nothing but an abnormal caprice he forfeits the least claim to mercy, since the act which seems noway essential to the preservation of his own life often endangers that of his victim: the wounds of bitten cattle sometimes bleed for days, and are apt to produce dangerous inflammations. The largest variety of the *Vampirus spectrum* measures nearly four feet across the wings, but is found only in Guiana. The smaller Brazilian species are very frequent; they are most troublesome in the darkest nights, and develop an almost miraculous instinct in the selection of their victims: in a roomful of sleeping people the soundest sleeper is always first attacked; and Baron Spix mentions the case of two drunken sailors who passed the night in the woods and were found almost *exsanguis* the next morning.

But, after all, the vampire-plague is a mere trifle compared with the Kalong nuisance: there is reason to believe that the myth of the harpies must have been derived from the winged gluttons whose countless swarms infest the forests of the Eastern Archipelago, and whose ravages would exceed those of the Egyptian locust if their habitat were not a region of inexhaustible fertility. The larger varieties are often brought to Holland; and an Amsterdam curiosity-dealer once showed me a pair of Javanese Roussette-Kalongs (*Pteropus vulgaris*), the only absolutely insatiable creatures I ever saw, though I have raised young caterpillars and hawk-owls. Night or day made no difference to them: the moment their box was opened they thrust out their fox-like heads and proceeded to gape with jaws that seemed to open by sections, revealing additional teeth in the far interior of the skull. Whatever those jaws could compass went down at one gulp; larger morsels were mangled rather than masticated, and in my presence the he-Kalong swallowed three pounds of boiled carrots in less than twenty minutes. Like maggots, bats seem to assimilate only a small portion of their food, as deglutition and excretion are divided by a very short space of time, and their voracity appears to be a vague desire to "fill up," rather than an appetite for any special kind of comestibles. Few soft organic substances of any kind seemed to come amiss to our Roussettes: potatoes, boiled meat, butter, bread, and bean-pods were devoured with equal greed, though not with the same rapidity as sweet fruits. By way of trying them, we once offered them spoonful upon spoonful of hashed beef, and, after gobbling about twenty ounces apiece, their swallowing process became somewhat laborious; but a slice of baked apple at once restored the vigor of that function, and they gaped as wide as ever. About an hour before sunset they began to get restless, and if the box was left open the he-Kalong would soon raise himself above the rim by means of his wing-hooks and move his head left and right, with an occasional grin of his foxy teeth. If supper was late, his mate would join him before long, and, after grinning and bearing it for a while, their impatience generally resulted in a quarrel: they would hook away at each other and utter their peculiar cry, a series of shrill whistles, varied only by prolonging or abbreviating the pauses. At the sight of a caterer they changed their whistle duet into a sort of twittering, and stopped it at the first mouthful, having now found a better use for their snouts; but if the visitor came empty-handed they expressed their disappointment in a curious way, by

A FOX-CHASE IN THE AIR.

dropping back into the box and scratching themselves violently with their long hind-claws. If that failed to propitiate the fates, they scrambled out and prepared to take wing: it was the hour when their Asiatic relatives get ready for business.

Near Cape Angol, on the southern coast of Java, there is a small mountain-village, Rydenberg or Rydenland-Koop, which has become a favorite pleasure-resort of the Dutch colonists, especially in midwinter, which here corresponds to the dog-day season of the Northern hemisphere. In spite of his phlegm, Mynheer is a keen sportsman and a remarkable shot, as certain neighbors of his had lately an opportunity to ascertain, and in the vicinity of Rydenberg large game is pretty well cleaned out; wild hogs are getting scarce, and tigers are now only found on Wynkoop's Bay, some forty miles farther west. Monkeys, however, are still plentiful, and all new-comers are treated to the favorite evening sport of the Javanese Boer,—a "fox-chase in the air."

Rydenberg overlooks the sea, and, some seven miles southeast, an archipelago of low islands, mostly well wooded, but uninhabited on account of their pestilential swamps. From these islands there comes in the evening a stridulous noise, resembling the distant cries of a sea-gull swarm, but shriller and wilder, and a few minutes before sunset large winged creatures rise from the jungle, mounting higher and higher in ever-increasing numbers, till the example of their leaders gives the signal to start for the coast. As they approach, their bird-like forms assume stranger proportions: zigzag-winged, and with heavy flops, they pass overhead, or plunge into the bamboo brake with an impetus that sways the tall stalks like reeds. Others fly along the coast toward the marshes of Wynkoop's Bay; but the plurality direct their course to the next fruit-plantations. The natives, however, are ready for them. Every farmer has from fifty to five hundred square feet of bast nets of all sizes and forms, roof- and funnel-shaped pieces for the orchards, and flat ones for the fields,—for the Roussette attacks corn- and melon-patches as well as fruit-trees. Judging from the ravenous appetite of the Amsterdam specimens, I should be inclined to credit the statement of a Batavian naturalist that a dozen Kalongs will strip a full-bearing plantain-tree in a single night,—*i.e.*, devour from sixty to eighty bananas in about seven hours. They cling to the fruit-clusters like parrots, skin a banana without breaking it off, and eat it down to the stalk in less than five minutes, and at once commence operations on the next one, often taking snap-bites left and right to ascertain the comparative maturity of the different clusters. Near Rydenberg, at an elevation of nearly three thousand feet, some tree-fruits need all the sunshine they can get, and the nets are therefore taken off every morning and replaced toward evening, which has the additional advantage of protecting the crop against the heavy thunder-showers which generally come down after sunset. If a fruit-tree is left uncovered, the Kalongs find it by the same unerring instinct that guides rats to an accessible granary, and the sportsman who ambuscades himself in the top of a guava- or mango-tree is pretty sure to sight his game before dark. Few other animals are so hard to kill and at the same time so easy to cripple as the *Luft-fux* ("sky-fox"), as the colonists call the large Roussette. The Javanese Kalong attains the size of a pug-dog, and in proportion to his weight his wings are just barely large enough, so that the least injury to his flying apparatus is sufficient to bring him down. On terra firma he tries to dodge behind trees and bushes the best way he can: finding escape impossible, he becomes aggressive, and attacks the boots and even the knees of the pursuer with his sharp teeth. I was shown a thick rattan walking-stick that had been bitten into splinters by a wounded sky-fox.

But to be fooled with nets or floored with lead is a sad alternative, and in wet years, when wild berries rot away before the end of the summer, the Kalong sometimes tries to circumvent the retiaries by turning out an hour sooner than usual, before the natives have secured their orchards. It is astounding how fast the hue and cry spreads on such occasions: men, women, and children seem to vie in giving the most audible proofs of their devotion to the public welfare. "Bhunderyak!" ("monkey-birds") yells the boy who was climbing a tree and happened to espy the harpies *in flagranti:* the laborers in the field, the women at the spring, take up the alarum, and soon a posse of villagers rushes forth with slings and stones, bent on revenge, the chance for prevention of crime being past: the sky-foxes have already settled on the seaward orchards, and may have stripped the best trees by this time.

The Kalongs know what is coming, and are all in a flutter, ready to decamp at a moment's notice, but still resolved to make the best of the remaining minutes, and eating away with might and main as they hover about the ripe clusters. At the sight of them the villagers approach with stealthy steps, till suddenly the stones begin to fly, pebbles as big as eggs hurtling through the tree-tops like a storm of grape and canister. Then a rush ahead,—the Kalongs have taken wing and are hurrying off seaward; but, even as they sail away in headlong flight, their ranks are decimated by smaller stones, and more than one sky-fox comes flopping down, flopping backward also in a desperate attempt to regain the shore, well knowing that in the water he will suffer a speedy sea-change in the maw of an Indian shark.

CHAPTER VI.

SACRED BABOONS.

SOME fifty years ago, the English naturalist Waterton conceived the idea of turning his paternal estate into an asylum for persecuted birds and beasts. He surrounded the entire domain with a stone wall eight feet high, and never allowed a shot to be fired on his grounds, in order to try "how tame kind treatment would make the shyest children of our All-Father." The wall, however, does not seem to have been high enough for the mischievous boys of the neighborhood; and Charles Waterton's pets never got rid of that hereditary dread of the bimanous species to which their ancestors had owed their safety for perhaps a thousand generations.

But the ideal which the British experimenter failed to attain has been fully realized in the birth-land of the human race, in Nepaul and Hindostan, and especially in the Ganges Valley, where the preservation of primitive habits and the doctrine of metempsychosis have made man the brother and playmate of his dumb fellow-creatures. Nearly all the South-Asiatic vegetarians treat mischievous animals with a more than Christian forbearance; but the worshippers of Brahm have, besides, been taught to regard certain species of the brute creation as half divine, and, consequently, altogether inviolate and entitled to the active charity of every true believer,—the most privileged of the zoological demigods being the bhunder baboon (*Papio Rhesus*), the Honuman (*Semnopithecus entellus*), the Brahmin cow, the pigeon, and the common crocodile. In Hindostan the public spirit of wealthy philanthropists rarely rises above the orthodox conservatism of the national mind; bequests are not devoted to public improvements, but rather to the maintenance *in statu quo* of incorporated societies and multitudes of secular and clerical mendicants; and Sir Emerson Tennent estimates that the produce of fully ten per cent. of all the stipends of a most charitable population of one hundred and sixty millions is consecrated to the support of lazy or mischievous brutes. The Dheva-Ghee, or purveyance system for necessitous animals, comprises some forty or fifty hospitals and several hundred food-dispensaries, some of them large enough to maintain a brigade of able-bodied Sepoys. All the Brahmin temples of the Bengal Presidency feed pigeons; many of them both pigeons and cows. Cows and monkeys enjoy the freedom of several wealthy cities,—are permited to camp in the streets and help themselves to whatever garbage and surplus fruit the market affords. Near Benares there are enclosed tanks where sacred crocodiles are fattened upon the meat-offal of the large city. The *mahakhunds* (literally, "big yards"), or monkey-almshouses, are found near every town and larger village throughout the Eastern presidencies; the honumans have special establishments where no low-caste monkeys need apply; sick and decrepit honumans and rhesus-baboons are tenderly nursed in several well-appointed hospitals that derive their resources from stipends or pious contributions. Wealthy Buddhists, as well as Brahmins, have often secured a local immortality of glory by the foundation of new mahakhunds, whose charters are sometimes codicilled with peculiar provisions: that vulgar monkeys and pigs shall be rigidly excluded from the benefits of the stipend; that such interdicts shall be suspended in years of famine; that the distribution of food shall always be superintended by a dhevadar of the charitable race of Sahib Jaghir Shing; that bhunder-baboons shall be entitled to two full meals a day, the surplus, if any, to be distributed for the benefit of pilgrims and low-caste monkeys, with the exception of the dancing macaques kept by jugglers and infidels; that legal fast-days must be duly observed, etc., etc.

Besides, the favorites of Brahm find a free lunch at the house of every true believer. A sacred bull must never be expelled from the enclosure of a truck-gardener without a fair compensation,—a sugar-turnip or a handful of dates. Honumans are rarely interfered with if they honor the premises of a native with an unexpected visit; their caprices must be tolerated as the dispensations of beings entitled to the most respectful deference, and unbelievers soon learn to consult their own interests by avoiding an open violation of that rule. It is far safer to thrash a Hindoo than to kick a sacred baboon; forgiveness of personal injuries is a duty, but all worshippers of Brahm will risk their lives in defending his favorites. A Hindoo offering violence to a sacred cow would be promptly stoned; an Englishman would be hooted, pelted, and before long probably waylaid and killed. If he could defy them, they would ostracize him, maltreat his servants, and secretly annoy him in every possible way. When Captain Elphinstone's Scotch gardener crippled a bhunder-monkey, the natives howled around the officers' quarters for fifty or sixty successive nights, besides carrying the baboon in procession and nursing him like a sick prince. In Bengal, baboons and crocodiles enjoy, in fact, all the privileges which the bigotry of our ancestors accorded to the monastic orders, common quadrupeds at least the prerogatives of a modern clergyman.

The results give a fair idea of the natural disposition of wild animals before their habits were biassed by the influence of the *Panic* emotion,—the terror which man himself may have experienced in the imagined presence of a mischievous divinity. Lizards do not dart out of your way, but just crawl aside to let you pass; a fish-hawk will alight on your gate and allow you to approach within ten or twelve feet before he betakes himself to the next tree. A sacred bull won't go out of his way to please the Governor-General. He encamps all over the sidewalk on the shady side of the street, letting saints and sinners take their chances in the gutter. The vegetable-market is his favorite stamping-ground; a little frolic now and then must be submitted to by the best of foreign residents as well as natives; when the reverend quadruped indulges in a frisk, the bipeds must pick up their bananas or bones and say no more about it. The sacred crocodiles bask on the shore and don't mind it a bit if you should indulge in an uncharitable remark about their plethoric appearance as compared with the condition of the human natives; but if one pelts them with pebbles they will turn their heads with a vicious snap, though without thinking it worth while to pursue a fugitive, their digestive powers being pre-engaged.

But the monkeys commune with their Darwinian relatives on a footing of equality which the Watertonian method would probably fail to establish in less than forty generations. My countryman Dr. Vanjorden went to Northern India as a scientific *attaché* of Lord Dalhousie's expedition, and during a residence of five years in the Punjaub and about three years in Bengal and Western Nepaul availed himself of several opportunities to visit the principal mahakhunds of those monkey-ridden regions. The baboon-asylum of Bhonaghir, near

Hyderabad, feeds about two hundred regulars and fifty or sixty occasional guests, the latter being visitors from the river-bungalows, whose summer vegetation they

THE PETS OF THE MAHAKHUND.

generally prefer to the somewhat arid neighborhood of the mahakhund. Breakfast is at eight A.M. sharp; but the dhevadar beats no gong: his boarders are sure to be on hand. The *menu* consists of rice, turnips, panicum (a sort of millet), pumpkins, and now and then a bushel of figs, served in a pile on the floor, between two troughs full of water. As soon as the gate opens, the guests crowd in, the old sachems first, the stout squaws a good second; but at the sight of any extras the press for precedence overrides all etiquette, and the dhevadar himself would be knocked down if he should presume too far on the deference of his *protégés*. They are on the watch for him if he enters the building, and when he reappears with a bucketful of tidbits they charge him with a rush, empty his bucket, clamber all over him in search of hidden sweets, and often use him as a jumping-board as they chase each other round the yard. In about four minutes from the first creaking of the gate the provisions are generally disposed of,—provisionally at least, Providence having provided each baboon with a cheek-pouch of such elastic capacities that a day's rations can be stowed away in one cheek. Lack of "cheek" is, indeed, no constitutional foible of the *Papio Rhesus:* he takes all he can get, and shares with nobody if he can help it. A fat old poucher, both cheeks distended with millet and his four fists full of good extras, will retire into a corner and growl viciously at the wistful look of a starved youngster. Woe to the low-caste monkey who should attempt to glean the crumbs of their feast! they charge him like bull-dogs, and somebody at the gate does not fail to take him across the knee and search his cheek-pouches before dismissing him.

By dint of much persuasion the doctor once induced the major-domo of the monkey-castle to postpone the usual time of the morning meal for an hour and a half; and the consequences fully justified the reluctance of the official. A few minutes after eight o'clock the young baboons became fidgety, and some of the elders, after strutting up and down in sullen silence, walked to the gate and began to shake the latch-handle, gently at first, but by and by with fierce impatience. Then, stepping back, the sachems held a council of war, chattering at each other with protruded lips, and grunting indignantly whenever they looked at the still unopened gate. Some of the frivolous youngsters were roughly handled, likewise the dhevadar's dog, who had mocked their grief with his ill-timed familiarities. Seeing a man approach from the direction of the village, they gathered around him, evidently in hopes of entering the gate in his wake; but when he pursued his way they vented their disappointment in howls that made the man stop and look back with surprise. After another consultation, they tried the gate once more, scrutinized the smooth masonry of the wall, and then made for a high tamarind-tree that overlooked the yard of the mahakhund. With some difficulty, but with grim resolution, the fat presbyters ascended to the very top-branches and began to challenge their landlord with louder and louder acclaims, rising at last to yells that must have been heard in the distant river-plantations,—for, soon after, a deputation of bungalow-baboons came hastening up the rocks, and joined in the chorus before they could possibly have ascertained the cause of the uproar. Boys, too, appeared on the scene; and when a fortissimo yell in the tree-top was answered by a shout from the village, the doctor himself advised the dhevadar to open the gate.

In another mahakhund, devoted to "honumans and pious and continent paupers," the guests, under a similar provocation, behaved with more dignity, though they, too, evinced a disposition to wreak their ill-humor upon the naughty youngsters. But toward low-caste animals the honumans show all the intolerance of the bhunder-baboon; sucking pigs coming within reach of their long arms are often grabbed and flung through the air with a suddenness that leaves the squealer no time for a protest till he lands sprawling on the other side of the fence. In hospitals for promiscuous animals race-prejudices are, of course, out of place, and under such circumstances even the honuman communes with his fellow-monkeys on more familiar terms, and often behaves with great kindness,—still, however, with a certain condescension, like a Church-of-England divine in the presence of dissenting ministers.

And the record of his caste seems to justify such pretensions. When Ravan, the Prince of Darkness, made war upon the Rishis, says the chronicle of the Upánishads, the monkey Honuman offered his services to the God of Light, and suggested the idea of carrying the war into the enemy's country by setting fire to the island of Ceylon. The success of this stratagem brought the Ravan party to terms and re-established the supremacy of the Rishis, but in the heat of the Ceylon *fracas-à-feu* the faithful ally's tail caught fire, and he would have expired in his own conflagration if he had not saved himself by a hurried trip to the Himalaya highlands, where he quenched the flame in a sacred mountain-lake, not, however, before his hands and face had got badly singed. The verity of this miracle is attested by the scriptural evidence of the Sâma-Veda, and, as a collateral proof, as our theologians would say, the honuman's face and hands are soot-black, and a tarn near the sources of the Jumna is to this day called the Bhunderpouch, or Monkey-tail Lake. Nay, the Buddhists of the Rayanate of Pegu in Ceylon claimed to possess an eye-tooth of the veritable original Honuman. It is an historical fact that in 1581 Constantine de Braganza, the Virey of the Portuguese colonies, captured this tooth, and that the Raya of Pegu offered him three hundred thousand cruzadas for the restitution of the sacred relic. The Virey hesitated, but his confessor insisted that the tooth must be destroyed, "as its surrender would abet idolatry, and probably witchcraft."

The piety of the Hindoo shrinks from all familiarities

with so sacred a creature; foreigners who wish to domesticate a honuman must treat him as a guest rather than a pet. Near the mahakhund of Khunar in the Nilgiri Hills there is a hygienic hotel where the garrison-officers of the Madras Presidency use to spend the hot summer months, and Dr. V. gave me an amusing account of the precautions by which the dhevadar tries to protect his saints from the irreverent tricks of the unbelievers. He feeds them early in the morning, before the luxurious Britishers have left their beds, and again at the very hottest hour of the afternoon, when sanitary considerations keep the foreigners within doors, and conjures them with prayers and lectures to shun the precincts of the hotel. His *menu*, however, is rather ascetic, while the heretics luxuriate in all the delicacies of the Madras market; and even saints have a foible for such dainties as pineapple jelly and preserved mangosteens. Dinner is at five P.M., and soon after the second gong the honumans put in an appearance, generally at the east side of the hotel, where a plantation of young myrtle-trees screens them from the observation of the dhevadar. The *Semnopithecus entellus* is naturally a frugal feeder, but the influence of an evil example is almost incalculable, and during the absence of the waiters (all Hindoos, though of doubtful orthodoxy) it appears that the favorites of Brahm were often induced to partake of flesh-food, and, as the dhevadar mentioned with bated breath, also of alcoholic beverages. The matter would have been less serious in regard to the neophytes of the flock, but the college of presbyters included an old grayhead with a milk-white tail,—an infallible sign of Jana-Ghitra, or canonical dignity of the fifth degree. And, grievous to say, this dignitarian was afflicted with an uncontrollable hankering after "jungle cocktails," a mixture of rum, sugar, and citron-juice, supposed to possess a prophylactic value in the treatment of jungle-fever. In vain did the dhevadar wrestle with him in prayer, in vain had he loaded him with amulets; nearly every Saturday night the whoops of a well-known voice from the direction of the hotel told him that the old man had been indiscreet again,—not drunk exactly, for as soon as the mixture began to take effect the waiters used to hustle him out. But one idle morning the officers found him prowling around the fence, and, guessing at the nature of his wants, took him aside and treated him to a bottle of *Nordhauser's Best*,—or *worst*, from a moral point of view. Before the waiters could lay hold of him, the dignitarian, bottle in hand, jumped out of the window and hastened to the esplanade, where the officers received him with cheers that soon attracted an astonished crowd of Hindoos and honumans. The animal of superior sanctity retreated to the top of a gatepost, and—but the details of the scandal are too painful to relate,—suffice it to say that two messengers from the mahakhund were so shocked at the impropriety of his conduct that they could hardly muster the courage to summon the dhevadar, who at last sent a peremptory order for all true believers to withdraw. The next morning the repentant saint, with his head thickly bandaged, was seen in the hands of a committee of Brahmins and Hakims, who nursed him with devotion, though they seemed to fear that his immortal part had been hopelessly compromised.

In the neighborhood of populous cities the pupils of a mahakhund are exposed to grievous temptations; pious visitors too often surfeit them with sweetmeats. The honuman-house of Kirni-ghar near Allahabad is burdened with a number of pensioners who are almost too plethoric to walk and seem to suffer all the horrors of dyspepsia. Such invalids are the objects of a special solicitude, their sufferings being considered as an illustration of the proverbial trials of the just; but in times of scarcity their lot becomes truly pitiful. Near Ghuyapor, on the lower Jumna, the scene of Krishna's dalliance with the milkmaids, Dr. V. saw the remains of a baboon-institute that had been abandoned during the late famine, and found the surrounding woods peopled with the ex-pensioners, now reduced to the sad necessity of working for a living, gathering berries and rolling logs and stones in search of coleopterous insects. The youngsters seemed to enjoy their occupation, but the old dyspeptics worked with groans, like the exiled aristocrats after the French Revolution.

During the Sepoy insurrection, too, the reckless

guerillas destroyed a good many mahakhunds, whose inmates were obliged to take refuge in the neighboring towns; and during the great famine of 1878-79, when the crops had twice failed throughout Bengal and the western Carnatic, bands of destitute monkeys roamed the country in quest of backshish, and were often seen around the dépôts of the Great Trunk Railroad, gleaning the offal of the grain-cars and appealing to the charity of the passengers. On the pike-roads holy honumans used to follow the palanquins at a trot, having found by experience that heretical travellers would sometimes feed them for the edification of the natives. In that time of great need many baboon-hospitals were abandoned, and even the jugglers had to discharge their dancing macaques, leaving them to pick up a living the best way they could. The poor things used to dance on the highway whenever they met a human being, and the *Benares Gazette* gave a touching description of a scene at the pier of the boat-bridge where two of the little Terpsichoreans waltzed around a blind beggar and every now and then approached him with beseeching squeaks.

An influx of high-caste monkeys has begun to gravitate toward the larger cities, for, considering the enormous extent of the country, the mahakhunds are, after all, few and far between, and the charity or the resources of the orthodox landed proprietors seem to have declined under the influence of the British dominion. But in the cities the Brahmins can still raise the wind to the pitch of fanning the fire of religious enthusiasm; and while there is flour in the barrel of a true believer there is always bread for the sacred baboons. Besides, hunger sharpens the wits of saints as well as of sinners, and with their four sets of long fingers the quadrumanous children of Brahm generally find ways of their own to keep body and soul together. They congregate at the river-wharves where the bumboats of the natives discharge their cargoes, and even canvass the European warehouses, though with more caution, on account of the sad irreligion of the low-caste Briton. In the market they mix with the crowd, and are apt to mistake spotted apples for offal and sound apples for spotted ones: after a fast-day they become semi-nocturnal and prowl around the stands of confectioners who sell their wares by torchlight. They also have a wondrous memory for faces and the localities of what our tramps call square-meal houses: a housekeeper who feeds a gang of baboons at the door of her residence can count upon permanent custom for the rest of the season. Subsequent rebuffs are unavailing; the saints yield to force, but come back the next day: true followers of their countryman Buddha, they seem to accept injuries as an earnest of benefits, and give the offender a chance to make amends.

Like Italian lazzaroni, city-baboons live in cliques,— clannish communities, very exclusive in times of scarcity,

SACRED BABOONS. 147

and always rather disinclined to enlarge their member-

FOUR-HANDED LAZZARONI.

ship except by natural increase and advantageous alliances, as with fat house-baboons of a roving disposition. Four-handed vagrants are promptly stopped and cross-examined: no mercy for the homeless stranger suspected of speculating upon a share of their scanty sportules, while the household pet with his brass collar and sleek pouch is merely scrutinized with silent envy. The half-grown bhunder-monkeys are so pretty that they are often domesticated, but their relatives dislike to part with them,—from motives that have nothing to do with "philoprogenitiveness." The holy children are their mediators, their apple- and bread-winners. The entreaties of the little beggars are not easy to resist: they will climb you after the manner of pet squirrels, embrace you with one arm and beg with the other, accompanying their gestures with a deprecatory mumble that becomes strangely expressive, as if they were pleading extenuating circumstances, if you offer to strike them. Even the idol-hating Mussulman is thus often beguiled into a liberality which his conscience may be far from approving. If the little spongers have struck a bonanza, they swallow *in situ* all they can find room for, well knowing that upon their return the contents of their cheek-pouches will be claimed by their relatives, for even a mother-monkey has no hesitation in plundering her own child in that way. To avoid coercive measures, the poor kids surrender their savings voluntarily and with great despatch at the approach of the ruthless parent. Like our artist-mendicants who keep a beggar-boy *ad captandum*, old baboons sometimes kidnap a baby of another tribe, keep a strict watch on its movements, but urge it with slaps and grunts to work the passers-by. Crippled baboons, too, are a most welcome acquisition to any clique. These twice-worthy objects of charity have their regular headquarters, where they can be found at any time of the day surrounded by eupeptic relatives who hope to participate in the largess of the pious. The poorest huckster will stop his cart in a gate-way to hand his tribute to a decrepit bhunder-monkey who supplicates him with outstretched hands. No true believer must stint his gifts upon such occasions; and so well does the hairy mendicant know the stringency of that duty that he flies out into a paroxysm of virtuous wrath if any passer-by should dare to disregard his appeal. The relatives promptly yield their aid, and fruit-carts are in danger of being monkey-mobbed if the driver hesitates to propitiate their resentment by a liberal contribution.

But the new-fangled conveyances of the foreign residents are thus often surrounded and stopped from sheer inquisitiveness. The Indian city-baboons have begun to take an abstract interest in human affairs. They will gather around a ranting quack, a revivalist, or a broken-down buggy, without any direct view to backshish. If a number of people run toward the scene of an accident, the monkeys race after them like dogs; if the Brahmins get up a pageant, the baboons join in the procession.

They take a curious delight in pressing their snub-noses against the shop-windows of the European merchants, and examine the array of novelties with a critical squint. A knot of strangers standing before a hotel, engaged in an animated discussion, has often been thrown into convulsions of laughter by the manœuvres of a honuman, joining them and chattering away with protruded lips and all the appearance of a personal interest in the issue of the debate. Fireworks, even long after sundown, never fail to attract a crowd of baboons, grunting their applause and looking at each other with approving grins. Housekeepers have to watch them carefully; for old baboons get very fond of toys. They will abstract a door-key, pick up a tin plate, a piece of brass, or an ornamental flower-pot, and run off with a demonstrative delight in their new plaything. A Delhi bhunder-monkey attracted general attention by parading the streets with two gaudy shawls, evidently not of legal acquisition, as his bad conscience made him take to his heels whenever anybody so much as pointed toward his drygoods.

Their long intercourse with the primate of their species has developed race-sympathies which often manifest themselves in an unexpected way. Colonel Lawrence, of the Agra "Planters' Hotel," keeps a tame leopard, which once followed its master to the freight-dépôt of the railway-station. The shady platform at the north end of the dépôt is a great resort for baboons and loafers;

and while the colonel talked to the receiving-clerk, his leopard strolled out to the platform, where a little street-Arab had fallen asleep upon a pile of gunny-bags. The moment he approached that pile a troop of baboons leaped upon the platform, and, instantly surrounding the boy, faced the intruder with bristling manes and menacing growls, evidently resolved to defend their little relative at the risk of their own lives.

But the trouble is that the Hindoos reciprocate such sympathies: the foreigners are strong and the natives weak, but we are few and they a great many, and experience has shown that it does not pay to hurt their pets. Lord Clyde ridiculed the idea of punishing a man for shooting a wild cow: it is now seven years in the penitentiary. Rough, no doubt, but a lesser evil than the revolt that would otherwise be sure to follow. The Sepoy insurrection originated in a quarrel of that sort: *beef*-tallow had been employed in lubricating the cartridges which the native soldiers were required to use. And in the eyes of a Brahmin a honuman is quite as sacred as a cow, and the crime of killing him (though less easily proved) quite as unpardonable. "*Bhara Nur!*"—" Mercy, mercy!"—is a frequent cry in the streets when a European domestic rushes out of a house in hot pursuit of a four-handed culprit. "*Sahib! Nenna san ghatta!*"—" We will make restitution, sir!"—they cry, if it appears that the sacred long-tail has got away with something; "hold! spare him for the sake of

Mahadeo! for Saki-Yam Deva's sake!" etc., etc., till the fugitive saint is around the corner.

"It isn't that rag I care for," said Dr. V.'s Prussian servant, whose neckerchief had been captured by a veteran honuman, " but the impudence of those fellows; I would give three months' wages if you would let me catch that old wretch and give him a Pomeranian *twenty-fiver!*"

A similar desire has got more than one Englishman into serious trouble. The naturalist Duvancel had to hide like a criminal when the rumor got abroad that he had killed and stuffed a young honuman, and, though he assured the natives that the deceased had met with an accident, the Brahmins appointed a committee to watch his garden day and night. Stuffing a honuman is almost as bad as killing him: the corpse must be embalmed and buried with due rites. The Frankish doctors are suspected of circumventing these regulations, and for that and similar reasons the city monkey-hospital of Benares used to be closed against all Europeans, with the exception of Lady Dalhousie, a recognized benefactress of the institution. Living monkeys are, of course, hard to watch, and harder to keep out of trouble, —wholesale trouble, sometimes, when the long-nursed wrath of an unbeliever explodes against them in some out-of-the-way place. But woe to the perpetrator of such a deed! an immortality of odium will be the fate of him who manages to evade temporal retribution.

"Wicked Harbarat's place" is, and always will be, the name of a certain estate near Agra, once the bungalow of a Captain Herbert, who had been tormented with honumans till he renounced the plan of turning the estate into a remunerative fruit-farm. But he retreated with a Parthian shot: the day after his departure some fifty or sixty martyrs, full of bananas and strychnine, were picked up in his garden.

Captain Elphinstone's servant, who had crippled a bhunder-monkey, was repeatedly pursued by a howling mob, and on one occasion was chased all over Delhi before he could give his pursuers the slip in the Mohammedan quarter, where a stout Unitarian kept the rabble at bay till the fugitive had effected his escape through a back-door. For the Moslems hate the baboons with an intense and perfect hatred, and, unlike the Franks, who are more apt to be reconciled by the comic features of the superstition, they denounce the monkey-worshippers as idolaters, outrageous provokers of Allah's threatened wrath. They post special watchmen to keep the hateful beasts out of their mosque-gardens; but, even there, expulsion and a kick *a tergo* is all they dare resort to: the pressure of public opinion is too much even for an Oriental fanatic. When the power of the Mogul dynasty was at its height, Shah Allum's Mahratta Peshwar (*Maire du Palais*) was once returning from his daily round of inspection when he heard that his youngest child had been attacked and viciously bitten by a troop of bhunder-baboons. The brutes had been captured *in flagranti*, and, as the chief culprit could not be identified, the incensed Mussulman sent the whole troop to the Selinghar, or state prison, which joins the royal palace of Delhi. The sentence was certainly not excessively rigorous, but before night the whole town was agog with ranting Brahmins and howling women. They kept up their lamentations all night, and the next morning, having ascertained the whereabouts of the martyrs, they shook hands with them through the grated windows and perfumed them with attar. The Peshwar was going to bring the matter before a municipal court, but the Shah induced him to enter a *nolle prosequi* and release the defendants.

In Agra, where the honumans are a terrible nuisance, the English Protestants have a cemetery of their own, and have come to the conclusion that the Sikh Lascars (discharged Mohammedan soldiers) make the only reliable sextons. The Rev. Allen Mackenzie was once summoned by a frightened messenger, who informed him that the "niggers" were going to gut the graveyard on account of some baboon-difficulty or other, but upon his arrival at the cemetery he found that the turbaned sexton had been already reinforced by an armed troop of his countrymen, who threatened to impale the first idolater who should presume to molest the faithful guardian of a *government* preserve. The whole fuss was about a couple of honumans who had been caught on the wrong side of the cemetery-wall and by their screams had attracted a swarm of two-handed and four-handed sympathizers. Some of the latter had taken refuge in the sexton's lodge, and when the mob had been persuaded to withdraw the irate official closed the lodge-door and attacked the intruders with a fury that defeated its own object, for the horrified

THE LIMITS OF HUMAN PATIENCE.

animals now burst through the windows and escaped with yells that came very near causing a new revolt.

Muhammed Baber alone was a match for the baboons. When they plundered his palace-garden he imprisoned them as fast as he could catch them, till the Brahmins volunteered to surround the garden with a high wall of smooth and absolutely perpendicular masonry. That is about the only remedy; for Brahm's favorites are too conscious of their immunities to mind a curse or the explosion of a blank cartridge. Human patience has its limits, and the holiest Brahmin would not see his last piece of bread snatched from his mouth without reaching for a boot-jack; but all fruits of our Mother Earth he would readily share with the eupeptic demigods. You may prevent the baboons by anticipation,— gather your fruit in time; but you must not expel the holy marauders, nor even forestall them altogether: pious farmers always leave a tenth of the grain-crop for the pigeons and monkeys. If a sacred crocodile takes a free lunch out of the calves of a true believer, it is guilty of misdemeanor, but it must be tried by its peers in holiness,—a court of true and accepted Brahmins. Unless the plaintiff prefers an indemnity, the sportive saurian may be found guilty, and is liable to be expelled from the stipend-pond. Under no circumstances must the layman take the law in his own hands; even secular magistrates have no competent jurisdiction in cases of that kind. On the cow-question casuists differ, but they agree that the animals must never be kicked out. You must try persuasion first, and gentle force only as a last resort. "Oh, my son, oppress not the poor!" Von Orlich heard a Hindoo farmer adjure a voracious bull. "Come, my child, I will feed thee with honey if thou wilt follow me." The bull continued to help himself. "Provoke not the weak," resumed the

SACRED BABOONS. 157

Hindoo. " Brahm is just; come, repent in time." The bull never budged, and the farmer at last summoned two companions. " Oh, my son!" they began again, but at the same time two of them seized the bull's horns left and right and thus trotted him out, chanting a passage from the Upánishads, while their assistant enforced the quotation by hammering a board with a sort of mallet.

Honumans cannot be disposed of in that way; you have to catch them first, and if you drive them over one fence the odds are that they will come back across another. They know their enemies, though, and keep a sharp lookout if secular reasons oblige them to visit the premises of an unbeliever. Only the brown face of a Hindoo encourages them to make themselves quite at home; and only the Hindoo farmer is ever treated to a full display of their gymnastic abilities. To see a swarm of honumans at play is a treat even for an East-Indian sight-seer familiar with the miraculous performances of the native acrobats. The evolutions of the boldest disciple of the Turner-hall would appear tame compared with the feats of the four-handed champion, for among the monkey-gymnasts of the Old World the *Semnopithecus entellus* has no superior and only one rival, the equally long-armed black gibbon. Haeckel seems to be right,—this earth must really be very old. Only the accumulated experience of many thousand generations can have developed such accomplishments. Without wings agility could hardly go farther; from the standpoint of a practical anatomist it is almost inconceivable how muscles and sinews, apparently so very similar to our own, can execute such movements. Without the least visible effort, the marvellous half-bird darts through the air in a wide zigzag, merely touching a branch here and there, upward suddenly with a series of mighty swings, regardless and apparently forgetful of obstacles, down with a gradationed spring that looks like a single leap, up again with a flying rebound through a tangle-work of branches, yet at the same time watching his comrades, aiming and parrying slaps or dodging a shower of missiles; then a sudden grab, a quick contraction of the hind-legs, and the acrobat sits motionless on a projecting branch, watching a movement in the grass that has not escaped his eye during his headlong evolutions.

The young baboons, too, make their summer life a perpetual circus-game, and if *panis* and *circenses* comprise the essentials of human happiness, the Hindoo farmer need not complain, and may, after all, enjoy his life quite as much as if he had exterminated the merry saints in order to save their tithe of the rice crop.

CHAPTER VII.

ANIMAL RENEGADES.

MORAL philosophers incline to the opinion that all the arts of Despotism have never yet succeeded in producing a perfect slave. Behind all the masks of non-resistance, under the thickest varnish of subordination, there is always a substratum of rebellious instincts; the love of independence is perhaps the most inalienable gift of Nature. It will re-assert itself after centuries of bondage,—even in brutes. No training and selecting has ever evolved a breed of absolutely domesticated animals; the tamest of them will now and then avail themselves of an opportunity to resume the life of their free-born ancestors. Household pets, that could not possibly profit by the change, have at least intermittent fits of independence. Only night-walkers know how much secret gadding our dogs are guilty of. On moonlight fields, on lonely mountain-meadows, one meets them, pair-wise and in troops, in quest of gallant adventures, but also singly, on strictly private business. Near the sheep-folds of the Southern Alleghanies sleek watch-dogs have often been shot as much as twenty-five miles from the homes they used to protect by their deep-mouthed barks,—till the inmates were asleep. Utter darkness, too, is apt to silence the voice of our faithful ally, and the next morning the people will wonder what makes the dear fellow so tired: the explanation might surprise them still more, if the Night could speak.

Domestic cats often absent themselves for weeks together, and return as lean as a rake, but unrepentant, till the dangers of vagrancy are brought home to them by boot-jack and gunpowder arguments. Many old village tomcats take regular summer vacations. Orchards and the extensive grain-fields of our Northern States supply them with young birds enough to keep soul and skeleton together, and the vicissitudes of roughing it seem to count for nothing against the pleasures of independence. The woods of the Mississippi Valley are full of half-wild hogs. They are just tame enough to answer a repeated dinner-call, but rarely come home of their own accord, though their adventures in the wilderness are rather over-spiced with danger: "bush pork" is generally full of buckshot. Goats, too, are apt to lose their way whenever they get a chance; and the hunters of the Tyrolese Alps often hear their bells in the inaccessible heights of the Ortler range, where they have to pick their food from the clefts of icicled rocks till the November storms drive them back to the valleys.

But where emancipation would be a change for the better, only constant vigilance can prevent a declaration of independence. In Eastern Europe, Southwestern Asia, and the Southern prairies of our own continent, millions of animals have permanently renounced their allegiance to the lord of creation. Wild dogs are not confined to the suburbs of Stamboul; legions of them infest the mountain-ranges of Armenia, Persia, and Turkestan, and prowl over the vast table-lands between Asia Minor and Northwestern India. They are found in the deserts of all intertropical countries; in America, especially on the arid plateaus of Peru, Paraguay, and Western Mexico. In Mexico and South America there are about sixty millions of wild horses and horned cattle whose freedom is bounded only by the limits of their speed. The *Cabras pardas* of the Sierra Madre are descendants of the Spanish goat, but as shy as big-horn sheep and nearly as hard to shoot. In all our Southwestern States there are utterly wild hogs, denizens of the river-jungle, and unapproachably shy. At the sight of a dog they stampede with snorts of horror and hide in swamps where few hunters dare to follow them, though the chase is perfectly legitimate. Their favorite haunts are the South-Georgian cypress-swamps; sporadically they are found as far north as Pamunkey Bay in old Virginia. The German hunters distinguish the *Wild-Katze* from the somewhat smaller *Feld-Katze*, the former the genuine wild-cat, the latter one-fifth smaller, and often with the fine fur of the ancestral tabby, but with all the fierceness of the genuine *Felis catus.* The alleged existence in the Ghobi Desert of a special kind of wild dromedaries (supposed to be the *Camelus primogenitus*) lacks confirmation; but there seems no doubt that the mountains of Balkh (the ancient Bactria) are the haunts of ownerless camels, that can be captured only by regular circle-hunts, for a month after birth their young ones are already too fleet for the dromedaries of the Bokhara nomads. The "wild asses" of the Old Testament, like the *Abu Ghibr* of Arabia Petræa, are probably survivors of a starved caravan, or deserters from the train of a defeated army, for in deserts where a horse would hopelessly perish his long-eared relative seems able to shift for himself; and Burckhardt asserts that the wars of Abd-el-Wahab have peopled the Arabian peninsula with herds of wild asses, resembling the shaggy Bulgarian variety. East of El Medina they roam in herds over the stony mountain-ranges, and generally give the city a wide berth, though in clear nights they pay an occasional visit to the pilgrim-camp of Bab-el-Musree to glean the waste provender of the caravans.

Near the precincts of the Eastern cities such four-legged independents are often merely domestic animals out of employment; but in sparsely-settled regions it is curious to observe the reappearance of their old race-habits. The *Khelp el Khamr* ("dog of the wilderness") of Asia Minor hunts in packs, and rivals his wildest relatives in the art of making night hideous with the true lupine ululatus, the long-drawn howl of his obstreperous primogenitor. In very cold nights they are apt to be-

BACTRIAN CAMELS.

come dangerous, and a few years ago the inhabitants of the vilayet of Khusabad rose *en masse* to avenge the death of an old sheik whom the Khelpies had killed and eaten in the neighborhood of a populous village.

The most interesting of my Mexican pets was a young *perro pelon*, or "tramp dog," whose mother had been imprudent enough to quarter her litter under the porch of a sacristia, or wayside chapel. But the child of the sanctuary had all the instincts of a young highway-robber. As soon as he could walk he waylaid the guinea-pigs and began to take a suspicious interest in the roosting-places of the landlord's chickens. The neighbors' boys brought him all the young rabbits they could catch, and he had a curious way of playing with them,—not like a sportive puppy, but like a young fox practising for business purposes. He would cripple them just enough to equalize the chances of the game, and then give them a fair race for their lives, taking care, however, to suppress any signs of excessive vitality. He never killed anything outright, but deferred his feast till incidental injuries had disqualified his victims for further sport. One half-grown coney, however, managed to get away from him, and would have escaped if the boys had not recaptured it; and when they restored it to him he massacred it on the spot, probably for having abused his confidence. Well-to-do house-dogs generally content themselves with eating their fill at the regular meal-times, but the *pelon* would never trust the chances of the next day, and invariably removed the remnants of his dinner, even potato-chips and tortillas. He had *caches* all over the farm, but especially in the rear of an old garden-wall, where he buried his bulky valuables; and the hogs that used to take their siesta near his treasury were always chased away and out of sight when he was going to make a deposit: he wanted no witnesses at such times. If I happened to surprise him at a grand interment, it was enough to make him nervous for the rest of the day: once in a while he would run back to the garden to see if I had not realized on my discovery. Of carrion he was so fond that he seemed to view the existence of his fellow-creatures from an ultra-Buddhistic standpoint, considering the speedy separation of soul and body as the chief object of their lives. Horses, especially, he regarded only as so many carcasses endowed with an annoying power of locomotion. He would often yelp at a big mare of somewhat frolicsome proclivities, eying her antics with disgust and with a mien of severe disapprobation of her frivolous delight in the vanities of life. The landlord's turkeys made him wag his tail; he was pleased at their fatness and the reflection that their vital propensities were far less incurable. The presence of man he accepted as a practical necessity, though perhaps with a secret leaning toward the view of the Encratian Gnostics,—that the removal of the bimanous species would at once restore the pristine glory of the globe. He seemed to "shun, not hate, mankind:" his favorite retreat was a gravel-hole beneath the old garden-wall, and nothing short of a fourteen-inch soup-bone would induce him to leave that place of refuge; appeals to his sense of duty were answered only by a stolid growl. I never heard him bark; his voice was an indescribable sort of half-howl, somewhat resembling the bay of a hound, though he used it rather as an expression of anger and pain. He was an incorrigible thief, and when the cook attempted to improve his morals with a broom-stick he transferred his headquarters to a neighboring mesquite grove, and finally evanished altogether, but continued to utilize his topographical knowledge, to judge from the frequent coincidence of dark nights with the disappearance of chickens and ducks. One evening I met him on the road to Fresnillo, and, recognizing my voice, he followed me as if nothing had happened till we reached the outskirts of the town, where he began to hesitate, and finally slunk off into a ravine, and that was the last I saw of him.

It takes several generations to eliminate the savagery of a "tramp dog." The Peruvian pampa cur (*Canis Azaræ*), though evidently the descendant of some domestic mongrel, is almost incurably shy. By dint of persistent kindness Rengger succeeded in gaining the confidence of a young pampa dog; but at the approach of a stranger he never failed to dart under his master's bed, howling as if he had a cramp in the stomach if the visitor so much as looked at him. The Mexican sierra-goats are less misanthropic and cannot be reproached with false modesty of any kind, but it is next to impossible to keep them near a farm. In winter-time they appreciate the advantages of a warm stable; but the advent of spring makes them restless, till one fine day they are off to the Sierra, sometimes in spite of wooden collars and drag-ropes. The kid-season, too, is apt to excite the migratory propensities of the dams; they do not like to bring forth in a land of bondage; some instinct seems to tell them that the Sierra is their proper home.

By a sort of spontaneous reversion, a similar instinct sometimes awakens in domestic pets; the mere neighborhood of a great wilderness seems to tempt them to desert. Among the wild cattle of the Brazos Valley the prairie-squatters often see a cow with a bell and an ornamental strap, perhaps the gift of a Missouri farmer's wife who advertised her pet as "strayed or stolen."

One of the most vivid recollections of my childhood is an encounter with the *bidet sauvage*, the wild pony that had roamed the Sambre highlands since the earliest memory of such little men as my companions. We were out after huckleberries, and had scattered among the high broom-corn and hazelnut-thickets of the plateau de Vence, when one of my comrades grabbed my arm and pointed toward a little knoll where a solitary horse was picking its way between the grass-fringed boulders. We crept nearer and nearer till we reached a ledge of cliffs on a level with the knoll, when my companion clutched me once more. "Go slow!" he whispered;

"oui, c'est lui, le bidet, the very pony: I know him by that stump ear. Stop! get down!"

We crouched behind the cliffs, but the pony had already seen us or somebody behind us: he started, stood still for a moment with his head high erect, then, leaping back with a snort, he wheeled around and flew over the plateau like a deer, down into a wooded dell and up the opposite mountains, where we saw him galloping along the ridge toward the head-waters of the Rouge-Air.

That same pony outwitted the hunters and herders of the Belgian Ardennes for more than eight years before he was finally shot near the Col de Grappe in Northern Lorraine. He seemed to know every pass and trail in the wide highlands, and even the favorite haunts of individual hunters; the game-keepers of Châteaumil had seen him more than twenty times, though never within shot-gun range and rarely without attracting his attention. During the hunting-season he was all suspicion and fled at the first echo of a shot, but in midsummer, when every wood was a hiding-place, he became more confident, and sometimes ventured into the lower valleys, where a cow-boy once saw him browsing peacefully among the parish cattle. The lad slipped away to summon his father, but when they came back with a musket the bidet was gone,—warned perhaps by one of those strange forebodings by which human outlaws have sometimes been saved from impending danger.

Upon another occasion a company of hunters had cornered him on a treeless ridge and opened fire as they contracted their circle, but when they had all but surrounded him he leaped down a cliff of twenty feet into the gorge of the Font-au-Loup Creek and disappeared among the broken crags. One deponent averred that he had watched him in the act of uprooting the bushes and weeds on a promontory he wanted to use as an observatory point; another had seen him drive a stray cow from his hill-pasture for fear that her absence would lead to a chase; and many other stories of that sort proved that we thought him capable of almost anything. That he was *bullet-proof* nobody ventured to question: it would have been an insult to all the foresters of the Sambre Valley. The antecedents of the old bushwhacker were somewhat obscure, but it was known that he had once been in charge of a farmer who kept a pasture for the saddle-horses of the Alleville hotel, and I suppose that the contrast between the green wilderness and the dusty pony-track so impressed his manly soul that he decided to secede. His forage-excursions were too well planned to get him into trouble, but at certain seasons of the year he was in the habit of visiting the lowlands on more risky business, and that habit finally proved his ruin. He thrice stampeded the mares of a large stock-farm, whose owner at last offered a prize of sixty francs for his skin. That started a hue and cry, and two weeks after the bidet met his fate in the form of a Lorraine poacher, who had seen him in the woods and who availed himself of that first chance to use his rifle upon legitimate game.

In a sparsely settled but tolerably fertile country animal refugees soon accustom themselves to the vicissitudes of their wild life. The ten months' drought of 1877, which almost exterminated the domestic cattle of Southern Brazil, was braved by the pampa cows, whom experience had taught to derive their water-supply from bulbous roots, cactus-leaves, and excavations in the moist river-sand. Solid food is only a secondary requirement; with a good supply of drinking-water many animals would beat Dr. Tanner's time. But how the Syrian Khamr dogs manage to make out a living only the gods of the desert know. They rough it in regions where no human hunter would discover a trace of game and where water is as scarce as in the eternal abode of Dives; nay, they multiply, for the Khamr bitch, like other poor mothers, is generally overblest with progeny: six youngsters a year is said to be the minimum. A sausage-maker would probably decline to invest in Khamr dogs: the word *leanness* does not begin to describe their physical condition; *strappedness* would be more to the purpose, if an Arkansas adjective admits of that suffix,—skin and sinews tightly strapped over a framework of bones. I saw their relatives in Dalmatia, and often wondered that they did not rattle when they ran; but Dalmatia is still a country of vineyards and

ANIMAL RENEGADES. 171

sand-rabbits, while the Syrian desert has ceased to produce thorn-berries. Without moisture not even a curse can bear fruit.

Where food is plenty, wind and weather seem to modify the *physique* of a tramp animal. Most wild dogs

MUSTANG COWS.

are bushy-tailed, gaunt, and fox-headed, and for some occult reason almost invariably *black-muzzled*. It is their clan-mark: judging from the snout alone, few naturalists would be able to distinguish a tramp dog from the pampa cur, the Khamr hound, the dog-wolf (*Canis Anthus*), or the Abu Hossein (*Canis Lupaster*). It does not improve their appearance; in connection with their wolfish eyes it reminds one too much of a hyena-head. Wild horses generally bear a strange resemblance to the ponies of the Russian steppe, and some of their characteristics may be recognized in the shape of the *mustang cows*, as the Texans call the half-starved cattle of the Mexican frontier. These horned mustangs, like their equine namesakes, are lean, knock-kneed, and thick-headed, besides having a rougher coat and a smaller udder than our domestic milch-cows. They are good fighters; their natural weapons resemble the terrible bayonet-horns of the Javanese wild cow and the more than half-wild *toros Galegos* that often turn the joke against the Madrid bull-fighters.

A singular character-trait of all animal renegades is their hostility toward their servile relatives. Travellers on the Rio Grande have to be very careful in picketing their saddle-horses, for if they stray into the prairie they are sure to be "mobbed" and cruelly kicked by the wild mustangs. A Bokhara courier, it appears, would rather meet a panther than a troop of wild camels; the mere sight of the gaunt monsters will frighten a dromedary out of its wits, and, unless the rider has much gunpowder to waste, the renegades, in spite of their timidity, come nearer and nearer, the cows stretching their long necks inquisitively, while the old males prance around with snorts that leave no doubt of their evil intentions. This rancor seems to be aggravated by a sort of *esprit de corps*, for in private life wild and tame beasts of the same species agree well enough and even pair, voluntary alliances between a dog and a female dingo, wild and tame hogs, mares and mustangs, etc., are by no means rare, but *en masse* their caste antagonism promptly asserts itself; just as a man may be the bosom friend of a partisan whose greeting in a public assembly he would hesitate to acknowledge: during the fever-heat of our sectional feud more than one dweller in Dixie thought it his duty to ku-klux his own brother. The only animal I ever saw torn literally into shreds was a Mexican butcher-dog that had followed us across the Bolson de Mapimi, the rocky plateau between the plain of Durango and the valley of the Rio Grande. The dog's owner, a poor Chinaco, had tried hard to sell him, but finally decamped with my partner's saddle-blanket, leaving his mastiff in lieu of payment; and, in accordance with a queer but well-known law of human nature, the poor quadruped then became the target of retributive attacks both verbal and practical; but, apparently mistaking our tent-wagon for the lurking-place of his missing master, he followed us with the resignation of a martyr. The Bolson is a *ravinous* country, and on the day after the Chinaco's departure we passed a precipitous gully at a place where a broken wheel and a lot of scattered boards marked the scene of a recent accident.

It looked like a slippery place, and, sure enough, down in the gully, some forty feet below the road, lay the carcass of a big mule, half buried in débris and surrounded by a swarm of tramp dogs. They had just begun their feast, and most of them were evidently in need of it: there were about twenty of them, two of the youngsters with a faint resemblance to half-grown shepherd-dogs, but all the rest of a more than wolfish leanness. Famine never reduces the body of a wolf beyond a certain point; his chest-bones make him look stout in spite of his starved belly; but the skeleton of a dog seems to shrink together with his bowels: some of the tramps in the gully looked as if their ribs had been strapped back upon their backbones,—"all legs and spine," like spider-monkeys. The shrinking of the lips had bared their teeth and gave them an unspeakably savage appearance whenever they leered at us with their deep-set eyes. Something or other seemed to excite them, and, looking around, I saw our friend the mastiff standing at the very edge of the ravine and looking down with a sort of pensive interest. "That's what folks come to who lose their masters," he might think to himself as he gazed upon the hungry tramps. But, while he gazed, one of the muleteers approached him from behind, lifted his foot, and in the next moment the mastiff's reflections were cut short by a kick that sent him head over heels through the air into the abyss below. What we call presence of mind is often nothing

ANIMAL RENEGADES. 175

but an instinctive impulse,—one of those instincts which a mortal danger awakens even in the human soul. Dogs are half human, guided partly by principles and prejudices, but in critical moments they act rightly from intuition. When the mastiff landed in the gully he picked

WILD DOGS.

himself up and stood still, rigidly still, facing the tramps, who had scattered in every direction but now gathered around him with ominous looks. They approached within ten or twelve yards and then came to a halt, watching the intruder with a steadfast gaze, silently, and with a gradual contraction of their haunches, like panthers crouching for a spring. Where the first movement is sure to be a signal of attack, even great strategists somehow prefer to let the enemy strike the first blow and thus betray his tactics,—" forewarned, forearmed,"—but circumstances are apt to disconcert such plans. A thing not larger than a hazelnut, a pebble thrown from the top of the rock, made the mastiff start just for a moment, but in that moment the pack leaped upon him with a simultaneous rush, and two seconds after the sound of cracking bones announced the end of the unequal struggle. They had borne him down at the first onset, and when they finally dragged him into the open gully I do not believe that there was an unbroken joint in his body. Three of the big tramps had done most of the killing, but now the whole pack laid hold, and in less time than it takes me to write the words they had torn him into pieces, not in the conventional but in the literal sense of the word,—limb from limb and rib from rib,—with a fury and a rage of destructiveness which plainly showed that hunger had nothing to do with their motives. It was evidently an act of revenge, provoked proximately by his unceremonious intrusion, but chiefly, without doubt, by the *odium invidiæ*, the pariah's deep-seated and long-cherished hatred of the privileged caste whose representative had dared to beard them in their den. What right had he to wax fat while they starved,—to fatten in the service of the arch-usurper of all the good things of this earth and then mock the leanness of virtuous liberals? "*La mort sans phrase!*"

Besides, dogs do not like to be interrupted in their meals, and a carcass-feast makes them especially touchy. I believe they are ashamed to be caught in an act of that sort; they seem to feel that there is something degrading about it. Carrion-eating is always more or less a last resort of famine: well-to-do quadrupeds leave such things to the maw-worms. The chief carrion-eaters are desert-dwellers, animals in reduced circumstances; for I am sure that even hyenas and jackals prefer fresh meat if they can get it. Vultures, on the other hand, have a natural preference for their ugly diet: I once caged a young *galinasso*, or Mexican king-vulture, and convinced myself that his cadaverous predilections were incurable. During my incidental absence he once remained a week without food or drink, and when I came back, having nothing else on hand, I gave him a young chicken, two handfuls of bread-crumbs, and a bowlful of water. He emptied the bowl to the last drop before night, but went to sleep without having harmed the chicken. They were together for the next four days, during which time the *gallina* ate all the bread, while the galinasso starved heroically; and when I killed the chicken he waited another twenty-four hours before he touched it.

The history of communistic insurrections shows that the chief wrath of the rebels is apt to explode against the tools of tyranny, while the sovereign can generally save himself by sacrificing a favorite minister. Four-legged mutineers, too, are mostly illogical enough to spare the Padisha of the animal empire, while they mob his pashaws. In stress of circumstances they recognize his superiority by claiming his protection; in America, especially, their independence has been too short to efface the traces of so many centuries of servitude. In Hindostan, where our black cattle come from, they are kept only for the sake of their milk and their sacredness; centuries before Herodotus visited the temple of the Egyptian god-bull, the Hindoos treated the cow as a privileged being, and it takes rather rough evidence to convince her that man is her enemy. The greatest North-American slaughterer of horned cattle is perhaps Captain J. Kellerman, proprietor of the Fronteras *matanza*, or beef-packery, near Matamoros, Mexico. He kills them and skins them by thousands, both at his establishment and in the open prairie, where his steeple-chasers wage unremitting war against all unbranded cows; but the survivors once proved that they trusted him, after all. He had pitched his camp near Aguaderas, in the midst of a big chaparral, when, just before nightfall, the crashing gallop of a cow-herd put his butchers on the *qui vive*. They made a rush for their horses, but there was no need of them: the cows headed straight for the camp, and by no means accidentally, for they only accelerated their career when they saw the camp-fires. When they had approached within a hundred yards, the captain saw that they were pursued by a troop of gray wolves, whose leader at last wheeled to the left about, while the cows kept right on, and, rushing into the camp, crowded, snorting and trembling, around the tethered horses. They were mostly cows and yearlings, some thirty altogether; and a Hindoo would probably faint to learn that the butchers "bagged" about twenty of them.

The Fronteras chaparral swarms with wild dogs, and during my stay in Matamoros the captain made a curious experiment with a "tramp bitch," whose puppies had been captured in the neighborhood of the matanza. The beef-packery is guarded at night by a dozen ugly-looking mastiffs, and the tramp dogs generally give the establishment an extensive berth; but in the hard winter of '76 they put in an appearance, at least in daytime, when the mastiffs were chained up. They used to sit in groups on the slope of a little hill near the matanza, appealing to the charity of the proprietor by yelping in chorus every now and then. There was so much waste stuff around the place that the captain concluded to grant their petition, and, by way of encouragement, sent them a car-load of beef-bones and "rippings," instructing the driver to scatter the scraps between the hill and the bone-pit. The tramps took the hint, and soon visited the pit every morning, in spite of the furious protest of the chain dogs. All went well for a couple of months: the tramps enjoyed their bonanza discreetly, and the chained mastiffs became hoarse and more tolerant. But in the horse-stable, behind the packery, a mastiff bitch had been quartered with her litter of puppies, and one evil day the door was left open, and the bitch at once made a rush for the pit. If she wanted a bellyful she missed her object, for the tramps killed and disembowelled her before the rescuing-party reached the scene of the conflict. Profanity is doubly heinous when it cannot mend matters: the bitch had been imported from Cuba, and her five little ones were all blind yet; but there seemed no help for them; there was no milch-cow on the place, and hand-fed puppies are a terrible nuisance. They were just going to drown them, when a Mexican boy-of-all-work suggested a better plan. He had seen a wild *perra*, a tramp bitch, that could be utilized as a wet-nurse. Whenever the perros entered the pit, she snatched up a bone and hastened back to the chaparral, and always in the same direction; once or twice she had come back within five minutes, so her lair could not be very far off. A promise of two dollars created a general interest in the enterprise, and before night the exploring party returned with the perra and eight *perritos:* they had tracked her to a hollow in a ravine and captured her with a common flour-bag.

Nursing animals do not like to adopt orphans while their own children are alive, and killing the perritos might make the mother still more intractable: so the matter had to be managed by stratagem. They chained her up in the stable and left her alone with her own puppies, but after an hour or so, one boy slipped a bag over her head while another substituted a young mastiff for one of the perritos, and so on, till she had five changelings and three legitimate puppies. The perra was as snappish as a trap-caught panther, yelled, howled, and made desperate attempts to break away; but the main point was reached,—she suckled the puppies, both her own and the mastiff's; nay, like the foster-mothers of young cuckoos, she seemed rather partial to the big substitutes. After a week or two her temper, too, improved; and when the puppies began to waddle around with open eyes she seemed reconciled to her captivity, as long as the youngsters did not crawl out of reach. But when they did, she often jumped after them with force enough almost to break the strap, and on one occasion not only almost, but quite, enough, for when the door was opened she darted out, and, clearing the fence with a single bound, whisked across the field and disappeared in the adjoining chaparral. She must have been very anxious to get away, for in the floor of the stable, close behind the door, she had dug a hole by tearing out a loose plank and excavating the stamped loam underneath, first outward and then upward,—so far up that another night's work would have liberated her anyhow. She had answered the purpose of her capturers, though; the puppies were a month old and had begun to eat alone: so the captain detailed a boy to feed them, and said no more about it.

The next morning one of the packery hands happened to pass the stable, and noticed a big hole, that seemed to have been dug from the outside in a way to communicate with a tunnel under the stable-boards. He informed the groom, his first impression being that the puppies were gone; the bitch must have fetched them during the night. But no; there they sat in their basket, all eight of them, munching away at some strange-looking object, which upon examination proved to be the body of a young *gazapo*, or mule-eared rabbit. There was only one possible explanation, though it seemed almost incomprehensible how the bitch could have dug a hole of that size in a single night,—a short summer night at that. And, moreover, how had she managed to elude the mastiffs? They had been unchained at sundown, and always patrolled the premises in every direction. The groom slept in the stable the next night, but nothing stirred; the night after, however, he was awakened by the yelping of the puppies, and, lighting his lantern, found that they were fighting over the remains of a big prairie-cock which some inaudible caterer must have brought them before midnight. It was now decided to recapture the bitch, if it could be done without hurting her, and the best plan seemed to be to catch her in her own trap by fastening a slip-noose over the entrance of her tunnel. But she was up to such tricks: five different times, at intervals varying from two to four days, did she visit the stable on her errand of love and get off safely ; only once the groom heard her scratch and fuss around, as if she had got into a tight place, but before he reached the trap all was still, and when he opened the door he thought he saw her skip over the moonlit yard. The lariat was drawn back into the hole, as if she had caught herself and slipped the noose off her neck. She always brought something or other, either game or a choice bone from the pit, and the puppies became so used to their nocturnal banquets that they whined all night whenever she omitted her visit.

The groom at last concluded to change his tactics. The stable had a loft with a separate door that could be reached by a rough-hewn stair of fifteen or sixteen steps. If the puppies were quartered in the loft, the bitch might try to reach them, and, finding the door locked, would probably dig and scratch, and thus awaken the groom. The plan was tried, and the puppies whined all night, but the perra returned no more. The love of liberty, after all, limited her maternal devotion, and within those limits she had done what she could.

CHAPTER VIII.

PETS.

THERE are instincts the study of which gives one a curious insight into the methods by which Nature attains her objects. Self-preservation is said to be her first law, and it is easy to see how "natural selection" could enforce compliance with such a decree: creatures that had mastered the art of taking care of themselves survived, the others perished; and the obvious necessity of that result still fills the school of life with eager pupils. But there are non-egotistical instincts whose real purpose has been carefully concealed. The inamorato blindly sacrifices his interests to those of the species. The ostentatious nabob becomes a patron of art and industries. "Vanitas," says Burton, "is a far better almoner than Caritas." The hobby-rider, the collector, the curiosity-monger, tug stoutly in the harness of science. Nature, it seems, rather mistrusts our sense of duty, and thinks it safer to bait a task with the semblance of a pleasure whenever she wants to engage our services on behalf of our fellow-men. .

With the same trick she overcomes the still greater

difficulty of employing the abilities of a superior species for the benefit of an inferior one. Against the resources of the constructive two-hander some of his poor fellow-creatures are unable to hold their own, and they would lean on a brittle reed if they had to rely on his Christian forbearance or on his recognition of their, perhaps somewhat recondite, usefulness. But the pet-mania solves the problem,—an instinct with an egotistical mask, but all its caprices shrewdly calculated to offset the effects of our destructive propensities. Helpless creatures can hardly be useful ones, but their dependence flatters our self-esteem, so we protect them, and Nature's purpose is answered. Finely organized animals need more care than others; we make them our special favorites, apparently on account of such incidental qualities as their playfulness and intelligence. We prefer rare pets, plausibly because of our fondness for out-of-the-way things, esoterically because they probably represent a species in danger of extinction. For instance, when the ur-ox, the ibex, and the bustard (*Otis tarda*) were on the point of being exterminated, they became such favorites with preserve-owners that their survival is now abundantly insured. There is a strange virtue in rarity. I suppose that our buffaloes, too, will become objects of *vertu* in time to save them from utter extirpation.

Curiosity-hunters sometimes dote upon creatures that would rather dispense with that honor; but, on the whole, protectors are in greater demand than protégés;

hangers-on are less often sought than found. Young animals are naturally submissive. The "myth-making propensity" of Monsieur Du Chaillu has perhaps been exaggerated, but I cannot help thinking that the stories about the uncompromising ferocity of his gorilla-babies must be apocryphal. The one that died last year in the Berlin aquarium-building was as playful as a child, and far more long-suffering and resigned,—"placid as a Hindoo," as Herr Behrens expressed it, and as, indeed, all analogies would lead one to expect in an animal whose anatomy, diet, and habitat are those of the vegetarian chimpanzee. Old orangs and chacma-baboons are churlish customers, but their young ones make most amiable pets; young tapirs, in spite of their pig-like stupidity, are by no means intractable; and I have often wished to try my luck with a young grizzly, for I am sure that jaguar-cubs can be made as tame as kittens. I raised one whose diet had certainly nothing to do with his gentleness, for I had nothing to give him but rats and beef; but I kept him nearly a year and a half before I ever knew him to hurt anybody intentionally; children and strangers could tease him with impunity, and I noticed that he always retracted his claws when the house-dog engaged him in a sham fight. The young of many animals, and especially of the feline species, have a curious way of parading their submissiveness by crawling to their master's feet, purring, and rubbing against his knees, or turning over on their backs,—a symbolic expression of unconditional surrender. They seem to feel their deficiency in useful qualities, and try to make amends by an appeal to our affections. The development of their natural weapons does not always awaken the disposition to employ them against a despotic master, unless circumstances assure them that his protection can

"JUANITA."

be dispensed with; captive baboons of an advanced age

still treat their keeper with a filial affection whose demonstrativeness fluctuates with the quality of the *menu*. For similar reasons danger often effects the sudden conversion of an infidel pet. The post-trader of the Fortin de San Pablo, near Mazatlan, is the nominal proprietor of an old ocelot that has long ceased to recognize his authority. Juanita absents herself for weeks together, and visits the post only as a guest, or rather as a privileged member of an inspecting-committee, for she rummages the premises, appears and disappears without asking anybody's leave, and resents every familiarity on the part of her former patron. But one evening she had just entered his store, when a troop of horsemen alighted at the gate, and a minute after a government scout with a big wolf-dog stepped up to the counter, while his comrades deposited their saddle-bags near the open door. Juanita cast an uneasy glance at the blockaded door, and in the next instant caught sight of the dog, and he of her, when the attitudes of both parties became so disagreeably suggestive of an impending set-to that the scout reached for a stick to chase his dog out. But Juanita either misconstrued his motive or had already made up her mind to secure a vantage-ground, for just when he faced about she leaped upon the counter, and with the next jump upon the shoulder of her old master, and there proceeded to "get her back up," growling viciously and bristling up into twice her natural size,

—exactly like a frightened kitten on top of an easy-chair.

Professor Brehm had a similar experience with a truant chimpanzee. The little scamp had the run of the Hamburg menagerie, and one day had managed to squeeze himself through the bars of the bear-rotunda, when one of the rightful tenants sallied from his den with a growl that made Jacko scramble up the centre-pole in wild haste. He found, however, that more than one could play at that game, for the bear espied him and came up the pole hand-over-fist; but when he had nearly reached him, Jacko jumped off, and clear outside of the enclosure, and then rushed into the arms of a by-stander, whom he hugged in a transport of tenderness,—"as a person saved from drowning would embrace his rescuer."

As a general rule, the spontaneous tameness of a creature depends on the degree of its helplessness, and the young of the most intelligent animals, being, with few exceptions, the least able to shift for themselves, are naturally the most anxious to secure a protector. Pigs can run and root almost as soon as they are born, and are remarkably independent cadets; puppies are cringingly submissive, and young monkeys not only accept but demand human protection. A young macaque, exposed in the middle of the market-square, will tackle the first passer-by, mount him, and cling to him as to a responsible relative, and fly out into a fit of exasperated jumps and screams if the stranger should decline the trust. One night I lost a little bonnet-macaque, together with a pet squirrel, and thought I had seen the last of them, as they had both been bitten by a savage cur whose owner had entered the garden-house by mistake. The squirrel had escaped to the woods, and never returned; but the next morning, as I was going toward the town, I saw my little macacus sitting in the middle of a cross-road, Micawber-like, waiting for something to turn up. The moment he saw me coming he made for me, but hearing a wagon approach from the other side, he turned around, jumped aboard, and took a seat by the side of the astonished driver. It was evidently not a case of personal attachment, but of philanthropy in general: like Madame de l'Enclos, he loved man *in abstracto.*

Billy Hammock, a mountain-squatter near White Cliff Springs, Tennessee, and supposed to be the champion fawn-catcher of his native State, informed me that most of his speckled pets had been caught by his little son in the huckleberry season,—*i.e.*, quite incidentally. It puzzled me how the little lad could have brought them home from the distant mountain-ranges he mentioned as his chief hunting-grounds, till he assured me that they *followed* him, after having been carried for a quarter of a mile or so; and, judging from the importunate tameness of an all but new-born specimen, I had no reason to doubt his statement.

The European stork seems naturally so fond of human society that he prefers the roof of a Dutch farm-house to the best nest-tree, and where he can be sure of good winter-quarters he has even been known to forego his yearly trip to the tropics, though his powerful wings would carry him in four days from North Holland to the rush-meadows of the Senegal. All intelligent birds can be domesticated, and the most intelligent of all, the common crow, is one of the few creatures that can be equally well tamed at any age. The old ones are harder to catch than any other birds of our latitude, but once boxed up they forthwith surrender at discretion, and in a day or two follow their captor all over the house and treat rival pets with vigilant jealousy. I have often wondered how tame crows and monkeys would probably be if they had been under civilizing influences for as many generations as some of our domestic animals,—chickens, for instance. The dawn-heralding cock is mentioned in the Sama-Veda; but sixty centuries of domestication have only half cured the innate shyness of his tribe. "Rushing around like a scared chicken," is an often-used phrase of the German language: corner a barn-yard fowl in a narrow lane, and see how it will illustrate the fitness of the simile. A tame crow under such circumstances would probably hop on your shoulder or step aside and let you pass. Anatomists could suggest one reason for the difference: in proportion to its size, a raven has about five times as much brains as a gallinaceous fowl.

The question whether there are any *untamable* animals requires a nearer definition of the somewhat ambiguous adjective. Untamable, in the sense of undomesticable, I believe there are none. With the proviso of a guarantee against socage-duty or a change of their natural habits, few animals would decline the hospitality of the *homo sapiens*, especially in countries where the sapient one has become the monopolist of all the good things of this earth. Let any one sweep the snow from his balcony, scatter the cleared space with crumbs, and put the balcony-key where the children cannot find it, and see how soon his place will become the resort of feathered guests,—not of town-sparrows only, but of linnets, titmice, and other birds that are rarely seen out of the woods. A little discretion will soon encourage them to enter the window and fetch their lunch from the breakfast-table,—by and by even in the presence of their host, for the fear of man is a factitious instinct, unsupported by the elder intuition that teaches animals to distinguish a frugivorous creature from a beast of prey. With so simple a contrivance as a wooden box with a round hole, starlings, blackbirds, martins, crows, jays, and even owls, can be induced to rear their young under the roof of a human habitation; squirrels, hedgehogs, and raccoons soon find out a place where they can get an occasional snack without having to pay with their hides.

Hamman, the famous German sceptic, used to feed a swarm of sea-gulls, often the only visitors to his lonely cottage on the shore of the Baltic. The neighbors suspected him of necromantic tricks, but he assured them that his whole secret consisted in never interfering with

STRANGE MESSMATES.

his guests,—keeping a free lunch on hand and letting them take their own time and way about eating it. The same magic had probably bewitched the pets of Miss

Meiringer, the daughter of a German colonist of New Freyburg, Brazil. Her father was a self-taught naturalist, and his collections have been described by several South-American travellers; but in the opinion of the natives his curiosity-shop was eclipsed by the menagerie of his daughter, who had tamed some of the wildest denizens of the forest, though evidently on the *suaviter in modo* plan, since most of her pets boarded themselves or only took an occasional breakfast at the fazenda. Among her more regular guests were a couple of red coaties, or nose-bears, several bush-snakes, and one large boa, a formidable-looking monster with the disposition of a lap-dog, for at a signal from his benefactress he would try to curl himself up in her apron, with a supernumerary coil or two around her knees.

There may be something, however, in personal magnetism. All menagerie-keepers know that there are persons who exercise over wild animals an influence which it takes others years to acquire. The chronicler of St. Renaldus tells a rather tough story about a troop of wild deer attending the saint's funeral; but the testimony of Moslems and Giaours seems to confirm the tradition that a *Santon*, or Mohammedan hermit, near Buda-Pesth had tamed the hill-foxes of the Bakony-Wald, and on his mountain-rambles used to call them from their burrows. Wordsworth's legend of the "White Doe of Rylstone" may also be founded on an actual occurrence, for that some attachments of that sort have had other motives than hunger and fear seems proved by many curious and often very circumstantial accounts of ancient and modern naturalists. Saxo Grammaticus speaks of a bear that kidnapped a child and kept it a long time in his den, and Burbequius, in his account of the Turkish embassy, mentions a lynx that had taken such a fancy to one of his men that his mere presence produced "a sort of intoxication" and his absence despair and finally the death of the animal. ("Legat. Turk.," chap. iii.) Pliny, the Roman Humboldt, mentions a tradition of a cow that followed a Pythagorean philosopher in all his travels; but where did he come across that strange story of the love-lorn *dolphin* that had been the playmate of a child, and, when the child died, came ashore in search of him and thus perished?

The tale of the Roman she-wolf, however, may be something more than a myth. In Dr. Ball's late work on Eastern Hindostan ("Jungle Life in India") there is the following curious account of two children in the orphanage of Sekandra, near Agra, who had been discovered among wolves. "A trooper sent by a native governor of Chandaur to demand payment of some revenue was passing along the bank of the river about noon, when he saw a large female wolf leave her den, followed by three whelps and a little boy. The boy went on all-fours, and, when the trooper tried to catch him, he ran as fast as the whelps and kept up with the old one. They all entered the den, but were dug out by the people with pickaxes, and the boy was secured. He struggled hard to rush into every hole or gully they came near. When he saw a grown-up person he became alarmed, but tried to fly at children and bite them. He rejected cooked meat with disgust, but delighted in raw flesh and bones, putting them under his paws like a dog."

The other case occurred at Chupra, in the Presidency of Bengal. In March, 1843, a Hindoo mother went out to help her husband in the field, and while she was cutting rice her little boy was carried off by a wolf. About a year afterward, a wolf, followed by several cubs and a strange, ape-like creature, was seen about ten miles from Chupra. The nondescript, after a lively chase, was caught and recognized (by the mark of a burn on his knee) as the Hindoo boy that had disappeared in the rice-field. He would eat nothing but raw flesh, and could never be taught to speak, but expressed his emotions in an inarticulate mutter. His elbows and the pans of his knees had become horny from going on all-fours with the wolves. In the winter of 1850 this boy made several desperate attempts to regain his freedom, and in the following spring he escaped for good and disappeared in the jungle-forest of Bhangapore.

Muhammed Baber, in his memoirs, speaks of a fugitive Afghan chieftain who was fed by a tame *mountain-wolf;* and there is no doubt that many pets of the larger species have voluntarily supported their owner instead of being supported by him,—especially where their employment agreed with their natural habits, though in animals, as in some human beings, there seems to be a certain *esprit*

THE GER-EAGLE.

d' office which in the service of an imperious master makes them do what they would not dare to do for themselves.

The last Rajah of Oude had a pack of hunting-panthers ("cheetahs"), that often took the field of their own accord, and used to deliver at least a portion of their prey, even if the expedition had not been successful enough to satisfy their own hunger. Nearly every Mexican cazique kept a trained eagle, whose value, according to Devega's chronicle, was often estimated at a sum representing the price of ten slaves.

That eagles can be utilized as well as falcons is proved by the experiments of the Förster Althofer, the overseer of an imperial game-preserve near Judenburg in Styria. He has trained both the Lämmergeyer and the *Steinadler* (golden eagle) of the Styrian Alps, but prefers the latter, and estimates that his pet ger-eagle saves him each year from twenty to thirty florins' worth of powder and shot. It is strange that the "gentle art of falconry" has gone so utterly out of fashion: on our Western prairies and in the water-fowl headquarters of Southern Florida it would be rare sport to slip a winged retriever, and, if the rush of our business life should leave us no time to do the training ourselves, we could get ready-drilled birds from Western China, where every landed proprietor keeps a pair or two.

But not only pretty or useful creatures find protectors; Vishnu has other pensioners on his list:

> For ugly things
> He findeth friends and food.

Some people seem, indeed, to select a pet on the principle that it is not likely to find other friends. St. Anthony's fondness for pigs may have endeared him to the hearts of his countrymen, but Lady Hester Stanhope's curs were such an eyesore to her Mussulman neighbors that they made wide détours rather than pass her home in the daytime. She kept leprous mongrels and tame jackals, as well as hunting-dogs. But her caprices were far surpassed by the eccentricities of Lord Rokeby, whose country-seat at Mount Morris seems to have been a promiscuous menagerie of the free-and-easiest kind. Dogs, pigs, monkeys, and young bears galloped up- and downstairs; a troop of fallow deer had their headquarters on the veranda and their parade-ground in the lower hall. The rooks had spread from the park to the turrets and garrets of the mansion, and defied the housekeeper, my lord being their helper. He, too, seems to have followed Hamman's plan of never touching his pets, merely giving them their board and their own way. Dr. Brehm's pet hyenas were long the marvel of his Hamburg fellow-citizens; but Frank Buckland's fondness for rats has been unjustly ridiculed,—they are really as playful as squirrels, and get wonderfully tame. Useless dogs are generally the most affectionate, and the same rule holds good of other animals: the most unprofitable pets are the most demonstrative in their attachments. Tame rats will lick your hands like little spaniels. Monkeys generally try to ingratiate themselves by entomological researches or by guarding the door against an imaginary foe,—listening and starting with well-feigned excitement, like a barking lap-dog simulating wheeziness by way of signalizing his official zeal. It may be real gratitude, though: a disinterestedly beneficent and, as far as they can perceive, omnipotent being must be a god in their eyes. Munificence charms even a quadruped savage,— unselfish munificence, at least. For it quickly alters the matter if we expect any services in return, especially such as involve a loss of personal liberty. In this sense a good many wild animals are not tamable, or only apparently so. It is curious after how many years of seeming resignation the involuntary recluses of our menageries will avail themselves of the first opportunity to escape. In the winter of 1875, Professor Rentz, the German Barnum, lost one of his lions during a freight-train collision near Fürth, on the Frankfort and Ratisbon Railroad. The deserter had been one of his performing animals, and during the last six years his keeper had often permitted him to leave his cage; but this had been his first chance for an out-door ramble, and he certainly made the best of that chance. Five weeks afterward he was shot near Villach, in Carinthia, having evidently tried to rejoin his free relatives, for in that interval he had travelled nearly a thousand miles southeast, or rather as nearly due south as the Alps would let him.

Monkeys can never be trusted in summer-time. The mere sight of a snow-storm is enough to scare them from an open window; but in the dog-days they cast many a wistful glance at the outer world, with its groves and apple-trees: man lives not by bread alone, and a velvet collar cannot reconcile him to a wire chain. Passive obedience is all one can expect from old-caught animals, and with those of the naturally pugnacious species it can be enforced only by a reign of terror. The wild representatives of the genus *Canis* will snap at your hand whenever you give them a chance. I once asked a German zoologist if there was no way of curing a jackal of that habit. "Oh, yes," said he: "measure him for his life, and thrash him twice a day within an inch of it." The devotion of the so-called pets of our travelling shows is often a sort of devil-worship: the panoply of the *Thierbändiger* ("beast-compeller") of Rentz's circus reminded me of the inquisitorial apparatus in the Nuremberg armory,—goads, nose-wrenches, leg-wrenches, spike-collars, hot-iron prods, pincers, choking-straps, whips, and knock-down clubs.

But there are pets that defy even such arguments. The most expert trappers of the Old World are probably the hunters of the Rhætian Alps on the border of Switzerland and Italy. They catch bears and foxes, kill thousands of squirrels, and visit the Swiss watering-places with cargoes of living pets boxed up in the smallest possible receptacles. Marmots form the staple of these peddling *cazatori*, but they keep also larger animals; and during my sojourn at Flüelen, on the Lake of Lucerne, a travelling *marmottier* sold my landlord a big mountain-lynx, warranted live and sound, though his temporary cage—a plank box with small air-holes—did not give him much chance to display his liveliness; but before the landlord paid the money he transferred the prisoner to a big chicken-cage of strong boards and faced with a door of stout woven wire. Darkness seems to cow wild animals, for in his new quarters the lynx soon began to snort around in a way that left no doubt of his warranted qualities: so the bargain was struck, and the Rhætian exile became a permanent boarder at the Black Bear tavern.

Some very ingenious bird- and fly-traps have been constructed on the principle that captive animals always try to escape lightwards, probably from an association of daylight with the outer air and liberty. For the first three days our pet concentrated all his efforts upon a certain corner of the door where the meshes were a little larger, and by grim perseverance actually succeeded in breaking one of the wires. But the only point thus gained was a sharp iron prong which lacerated his jaws in a frightful way, till the landlord pried the wire out and replaced it by a ten-penny nail. The prisoner then changed his tactics. Somehow the conduct of his jailers had led him to infer that their object was not to eat and skin him, but to retain him *in perpetuo* for his supposed amiable qualities: so he concluded to make himself as disagreeable as possible.

He would double himself up in a corner of his cage, looking unutterable things, and as soon as anybody came near the door fly at, or at least in the direction of, his face with an impetus that bent the wires, suggesting dire consequences if ever the door should give way altogether. These demonstrations he accompanied with a peculiar yell, something between a hiss and a howl, and in the night-time he often uttered that same cry, at uncertain intervals at first, but afterward with the regularity of a minute-gun. The guests complained, and the Bear landlord resolved to silence the serenader. He procured a big horse-syringe, filled it with absinthe, and made the hostler conceal himself behind the cage. Whenever the prisoner raised his voice, the hostler raised his syringe and drenched him with wormwood extract till he could not doubt that his laments only increased the bitterness of his situation. But despair is as inspiring as hunger, and somehow the lynx found out that the wood-work was the least impenetrable part of his cage,—nay, that the rear board in particular was of a less obdurate texture. This board he now attacked with tooth and nail, to which he superadded a concussive force by stepping back every now and then and leaping head foremost against the centre of the panel. The landlord watched his manœuvres, and finally got uneasy. "He's a Grison," said he, "and his countrymen are a headstrong set. We shall have to forestall him." He rummaged his garret and found just what he wanted,

—an old oven-grate, that converted the imperilled board into a heavy-barred window with a wooden shutter.

But the Grison found a road to freedom in spite of iron bars. He retreated to the rear of his cage, with his face toward the darkest corner, and thus remained motionless, day after day, though the disappearance of his provisions seemed to prove that he must spend his nights in a less pessimistic way. His serenades, at least, had never been resumed, and the landlord flattered himself with the hope that he was going to accept the situation, when the hostler discovered that his last two weeks' provisions had been hidden under the straw, and that the prisoner was *in articulo mortis*, to judge from the glassy appearance of his eyes, and from the feeble groans which the cover of his straw couch made almost inaudible. "I'm up to that game," laughed the landlord. "I had a fox that tried that same trick on me. We'll soon make him eat: all we have to do is to chuck out his straw; if he sees the meat, he won't resist the temptation."

But before the cage was opened the groans became lower and lower and finally ceased, and when we removed the straw we found that the Grison was already beyond worldly temptations: he had solved the problem of Gautama in a way of his own.

Miracles usually end where the Age of Reason begins, and it has been pointed out as a suspicious circumstance that snake-charmers are the almost exclusive product of semi-barbarous countries. But Dr. Grotius reminds us that the discovery of a new law of nature would enable any man to work apparent miracles; and there is no doubt that the out-door life of such long-headed barbarians as the Chinese and Hindoos has put them on the track of some useful zoological secrets. Observation and invention are two widely-different faculties; the crude empiricism of our forefathers has led to sundry discoveries which our analytical methods might have failed to achieve. Besides, the East-Indian beast-charmers belong to a special caste, a corporation that has carefully preserved its *trade secrets*. For, in Hindostan, snake-charming is a branch of a regular business that includes horse-breaking, rat-catching, monkey-training, and other occupations that must give their adepts a peculiar insight into the faculties and foibles of animal nature. A tame snake is only a side-show of the *Guruwalla*, or travelling vermin-destroyer; or, so to say, a living diploma of his mastership. He uses a trained cobra as a business-advertisement.

Jugglers with dancing snakes are seen on every Bengal market-place, but the grand masters of their craft exhibit very different tricks. Lord Dalhousie's guru-walla-en-chef used to call rats from their holes in broad daylight, and had a water-snake that followed him like a dog and could not be driven out of the room as long as its master was present. He had evidently established some hold on the affections of his strange pet, for he never failed to bring her back by a single whistle after flinging her into the middle of a pond where she could easily have escaped if her subjection had been an involuntary servitude. His favorite trick was to get a lot of the common black tree-snakes (*Coluber dryas*) and by a mere word make them engage each other in mortal combat,—snakes which but a minute before had lain coiled together in a sluggish repose from which no other human voice could rouse them even for a moment. By a slightly different sound he would inspire them with a panic that sent them darting out of the room, and out of the house if the doors were open; nor did he trouble himself on such occasions to recapture them, for, after a few hours' manipulation, a batch of fresh-caught snakes would serve his purpose as well.

The professional jugglers prefer the *cobra* only as the least expensive of all sensational animals, for crocodiles and pythons are equally tamable. The wardens of the sacred crocodile-ponds near Benares keep their pets under perfect control, and, as Captain Godwin Buchanan assures us, through influences among which hunger is only a minor item. His opinion seems confirmed by the statements of a famous Spanish beast-tamer, proprietor of a cockpit and pulque-shop in Tampico, Mexico. The effluent canal of the Tampico Laguna is well stocked with alligators, whose services to the health-police have made them a sort of public protégés. They live upon the offal of the slaughter-houses, and are so well fed that they can afford to spare their two-legged fellow-citizens:

they mind their own business and give bathers a wide

THE ALLIGATOR-CHARMER.

berth. But the cockpit landlord has a negro-of-all-work

who for a couple of coppers will convoke the caymans as a farmer would summon his pigs,—nay, often without any audible signal, by merely going to the water's edge and standing with uplifted hands till the alligators throng around him in crowds. He declines to divulge his *modus operandi*, but his employer is positive that he never feeds or touches his pets. About his private theory the *pulquero*, too, is somewhat reticent; but when my former colleague, Dr. Landgrebe, of Tampico, once asked him a home question,—"What could possibly induce the caymans to gather around a person who never feeds them?"—"*No sé*" ("I don't know"), he replied: "*se cogan los castores con el rastro*" ("beavers are baited with a scent").

In Europe the rat-catching business is monopolized by the gypsies, who may have imported their methods from their native country, for it is now an established fact that their race are the descendants of a tribe that left Hindostan during the reign of the first Mogul dynasty. In Austria, where the *zigeuner* are as frequent as tramps in New England, a rat-catcher will take a contract to expurgate a farm for ten kreutzers (about eight cents) a house, and twenty kreutzers the whole premises; and he certainly earns his fee. He uses both traps and poison; but the peculiarity of his bait is its instantaneous effect. With poisoned cream-cheese a man might kill a good many things in the course of a year; but the zigeuner will lock himself up in a stable, and after an hour or so come out with a bagful of mice, live or dead, at the option of his employer. Their incantations are confessedly a "blind," a sham imitation of an art which the masters of their guild reserve for themselves, for the exploits of the Oriental jugglers leave no doubt that musical instruments form the essential tools of their trade, and I have often wondered if the story of the Pied Piper, and even that of Orpheus, may not be something more than an allegory. The undoubted ability of the professional guruwallas to convene a troop of rats in broad daylight is not a whit less marvellous: the tricks of the crocodile-wardens might be founded upon a character-study of individual saurians; but the exorcism of a swarm of wary and timid domestic parasites seems to imply the discovery of a key to a generic peculiarity of such creatures.

Much less miraculous is the tameness of an old household pet, no matter of what species, for the daily intercourse with human beings has an almost incalculable effect in transforming the character of a captive animal. Next to the love of liberty the love of life is certainly the master-instinct of every living creature; yet domestication has subordinated this instinct to the wholly artificial sentiment of man-worship. At the bidding of a revered master, dogs, gerfalcons, horses, camels, and even elephants, will rush to certain death, and not blindly, either, but open-eyed and with a persistent suppression of the *horror naturalis* and a disregard of well-known dangers that must be called a deliberate self-sacrifice. Major Keogh's old roan, the only survivor of the Big-Horn massacre, was found limping about the battlefield with eleven bullets in his body, yet every now and then hobbling back to the place where his rider had fallen; nay, during Mehemet Ali's campaign against the Wahabees a troop of baggage-camels broke away from their captors and followed their comrades through the fire of a burning village.

Conscience, too, in one sense of the word, is, properly speaking, factitious instinct: as a synonyme of remorse it implies a *post-facto* feeling of compunction,—a feeling unknown to the creatures of the wilderness: instinct-guided, they act in conformity with their only standard of right, and have nothing to reproach themselves with. But the artificial circumstances of domestication alter that standard, and the instincts of a captive animal may betray it into actions which on second thoughts appear to be at variance with its true interests. Where a fox has once robbed with impunity he will try to rob again, unable to realize to what degree his actions may provoke the resentment or sharpen the wits of the injured farmer. If a mischievous puppy is not punished on the spot, it will expect to go scot-free. But an old dog knows that the prerogatives of man include the faculty of nursing his wrath. I knew a pointer bitch whose contrition quite disqualified her for business for the rest of the day whenever she had been guilty of a mistake. The herders of the Transvaal often leave their flocks in charge of the dogs, and upon their return to the pasture perceive at once if anything is wrong, if a sheep has been crippled, etc., for at the sight of his master the responsible dog will break out into a howl of abject terror. But the most curious instance of the power of conscience in animals is recorded by Professor Schomburgk in a communication to the *Bilder aus dem Thierleben*. He had taken charge of the zoological department of the Adelaide City Park, and was almost nonplussed by the inveterate mischievousness of a female bhunder-monkey. In solitary confinement she would alarm the neighborhood with her piercing shrieks; but the name of the happy-family cage became a misnomer whenever she was restored to the companionship of her relatives. Not content with teasing the young macaques, she would aggravate the old ones in every possible way, and had a dreadful talent for raising a general row; but thus far her offences had always been condoned by the intercession of her old keeper. One evening, however, this same keeper appeared with his arm in a sling and stated that the bhunder had tried to bite his hand off. His arm was dreadfully mangled, and the laceration of the wrist-sinews made it doubtful if the man would ever regain the full use of his hand. Schomburgk at once ordered the bhunder to be shot. Early the next morning one of the assistant keepers loaded a shot-gun to execute the sentence. The monkeys were quite familiar with the sight of this gun, which had frequently been used to shoot the rats that infested the premises, and when the keeper approached the cage they continued their several occupations with perfect unconcern. But with one exception: the moment the bhunder caught sight of the fateful implement she came down from her perch like a shot and darted into a back room,—the sleeping-apartment of the cage. It was near the breakfast-hour, and the keeper bided his time. Breakfast came, and the monkeys charged in like a lunch-brigade. But not the bhunder. Contrary to all her habits, she kept out of sight till her comrades had picked out the tidbits, and only when the keeper had stepped round the corner she slipped out, grabbed a piece of bread, and rushed back into her hiding-place. The keeper then tried a stratagem. The door of the sleeping-cage could be shut with a spring-bolt, and, after connecting the spring with a long strap, he posted a boy in the opposite corner of the hall while he crouched down below the platform of the main cage. After a full quarter of an hour, he saw the boy raise his hand, pulled the strap, and heard the door shut with a click. He now had the bhunder at his mercy, and her behavior showed that she knew it. When he pulled the strap, she made a rush for the door, and, finding her retreat cut off, began to "rage around like a wild-cat possessed," up and down the cage, with piercing screams, while her companions eyed her with mute astonishment. Finally, feeling her strength fail, and seeing no possible way of escape, she flung herself into a corner, where a much-deserved fate at last overtook her.

Intelligent animals rarely resent the severity of a trainer who once has made them feel his power; but that their forbearance must require a great deal of self-control is proved by the fact that they sometimes revenge themselves upon a proxy of the tyrant,—his friends or a favorite pet. After a knout-drill some hunting-dogs have an ugly way of falling upon their comrades, or even upon their own puppies, resolved to "take it out" of somebody. Sick horses often kick the stable-boy by way of getting even with the farrier; and I remember an amusing instance of an animal's appeal to the code of the *lex talionis*. During the winter season the Botanic Garden of Brussels is used by the proprietors of various peripatetic menageries as a zoological dépôt, where the caged travellers can recuperate and enjoy the hospitality of the city on condition of exhibiting their charms gratis. Sick animals often stay the year round; and a few years ago the managers took charge of a baby elephant whose constitution had all but succumbed to the rigors of the climate. In the course of the summer, however, Micheline got on her legs again, —so much, indeed, as to become positively rampant, especially when her keeper indulged her in an out-door ramble. On account of the supposed sensitiveness of her lungs she wore a woollen *couvrette*, or shawl-saddle, to which for some reason or other she had taken such a fancy that she would readjust it herself whenever it slipped down. But one morning she sauntered toward an open gate where the laborers had unloaded two big vats full of pickerel-spawn, and, finding the mixture pleasantly cool, she upset one of the vats and began to welter like a pig in a puddle. She had just upset the second tub when the enraged gate-keeper fell upon her with a cow-hide, and after belaboring her till her grunts changed into pitiful squeals, he snatched away the soiled couvrette and dismissed the culprit with a fifty-pound kick. Micheline had not offered the least resistance, but when she walked away she uttered a series of peculiar gutturals, sounding almost like muttered threats. She walked toward the orangery, and one of the gardeners who had watched the rumpus from a window of his lodge then became the witness of a curious scene. In the orangery the gate-keeper's children were at play among the trees, and, without the least provocation on their part, Micheline suddenly charged them, and, singling out the biggest boy, began to thrash him with her trunk just as the old man had thrashed her with his cow-hide. After dodging left and right between the bushes, the little lad ran screaming toward the gate; but the superior speed of his pursuer obliged him to take refuge in a tree, and before he could clamber out of reach Micheline grabbed his breeches—a worn-out pair, luckily—and tore them off with a single jerk. When the pitchfork brigade rushed to the rescue, she was strutting up and down with her trunk proudly aloft, waving the *spolia opima* over her head.

A DANGEROUS PLAYMATE.

But only hunters can realize the influence of education in controlling the passions of an impulsive animal. The Mongol Tartars hunt with trained panthers ("cheetahs"), and Kohl assures us that hunger itself will not tempt these cats to tear their prey before the arrival of the hunter. After the return of the Saracens from the Persian conquest the cattle of friend and foe got mixed, and the Commander of the Faithful is said to have identified the Arabian horses by the following test. He kept them three days without a drop of water, and then let the slaves drive them toward a river-bank. But in the moment when they saw the water and rushed ahead to quench their thirst he ordered the trumpeters to sound an assembly call, and one-third of the famished beasts actually wheeled around and galloped back to the camp. The word *ennui* does not begin to express the misery indoor life must inflict on dogs whose souls, like the Scotch exiles', are roaming through the Highland fells. But how resignedly do they await the pleasure of the complacent master who beguiles his leisure with page after page of printed adventures which his dumb companions can enjoy only in their dreams! No words can be more eloquent than the occasional inquiring look of a hunting-dog, sick with hope deferred, but whose only protest against martyrdom is his unbounded joy at the termination of it, when his master at last reaches for his hat and takes down his shot-gun.

The Hindoo fakir who fills his mouth with gall *in majorem Dei gloriam* cannot suffer more for Buddha's sake than many a town dog has to suffer in the service of a master who keeps a tan-yard or a chemical laboratory.

To a creature whose nose can distinguish the " cold trail" of a rabbit at a distance of sixty yards, odors which offend even our blunt olfactories must be as irritating as the continuous screech of a steam-whistle would be to the human ear or the sound of a fiddle to the ear of a bat. The upper story of the Salzburg Acropolis is infested with innumerable horseshoe bats, and the steward often uses them for a curious experiment. He claps one into a wire cage, puts the cage on top of a desk, and on a lower shelf of the desk a *Hackbrett*, or Styrian zither. At every twang of the zither the bat will start as if a fine needle had pierced its body, and a prolonged performance will throw it into a fit, a convulsive twitching of the whole flying-membrane. This same nervous twitching I sometimes believe I recognize in the grimaces of a town dog averting his head with a sort of shudder or rubbing his nose against the ground. Life would be a curse to some dogs if nature had not mitigated their martyrdom by blunting their senses. The effluvia of the sheep-fold have made the shepherd-dog almost scentless, though his form most unmistakably betrays his descent from the sharp-nosed jackal.

Arthur Schopenhauer maintains that the development of artificial faculties weakens our natural instincts; but it is likewise true that in lieu of lost instincts our domestic animals have gained several new faculties. If domestication has spoiled the nose of the average house-dog, it has certainly improved his ear. Dogs and horses have but scanty means for expressing their emotions, but their power of apprehending spoken words and other sounds far surpasses that of the parrot. A cavalry-horse learns to distinguish about fifty different commands, besides their equivalent bugle-signals. People who will content themselves with looked and acted answers can carry on a regular dialogue with an intelligent dog. A poodle will distinguish an exclamation from a command, a question from an invitation, a compliment from a persuasive coax, a warning from a taunt, and even a bantering taunt from a real reproof. The memory of an old hunting-dog is stocked with a regular glossary of venatorial slang, and the inability of animals to discern the elements of articulate speech only increases the wonder: they seem to depend exclusively upon the differences of intonation which a speaker somehow adapts to the sense of the essential words. Domestic pets will recognize their master in almost any disguise, but it is still more difficult to deceive them by a dissembled pronunciation: in the darkest night dogs and monkeys identify an old acquaintance by a single word, or even by the mere sound of his voice.

It is a strange fact that in night-time an unknown sound will scare monkeys almost out of their wits. The creaking of a wheelbarrow, a whisper, the rustling of a window-curtain, is enough to throw them into a fit of horrified screams and contortions; capuchin monkeys rush wildly through their cage, macaques try to force their prison-doors, the little marmosets huddle together like the princes in the Tower, all about a matter they would disdain to notice in daytime. The old males of the anthropoid apes are about as hard to scare as any living creature, but after dark the veriest trifle will inspire them

SURVIVAL OF THE FITTEST.

with an almost supernatural fear; and it may be a mere fancy, but I cannot get rid of the notion that this night-horror of our hirsute relatives must be the origin of the spectre-dread of savage nations, and indirectly, perhaps, of mediæval demonism and modern spiritualism, —"ghost-mongery," as the sceptical Germans call it. Monkeys are not very sharp-scented, and have to rely on their eyes, and in night-time, therefore, are almost at the mercy of their enemies, jaguars, panthers, and leopards, whose owl-eyes enable them to hunt by moonlight, and in the virgin woods of the tropics the constant dread of mistaking the approach of a murderer for the rustling of the fitful night-wind would be enough to make a Berserker nervous. "It is not books or pictures," says Charles Lamb, "nor the stories of foolish servants, which create these terrors in children. They can at most give them a direction. The stories of the Chimæras and Gorgons may reproduce themselves in the brain of superstition, but they were there before. They are transcripts, types: the archetypes are in us, and eternal." May it not be that those archetypes are the prowling *feræ* of the tropical forests?

There is a story about an ex-railroad-conductor who, in the fever-dream of his last disease, called out the forty stations of his route, in due succession, and at correct intervals, and the fortieth at the terminus of his life; but the power of habit manifests itself quite as strangely in the "second nature" of our domestic animals. The *trapiches*, or cog-wheel mills, of the Mexican planters are turned by horses, which have to make several thousand rounds in the course of the day; and in the solitudes of the chaparral it is nothing uncommon to see a revolving object which upon nearer investigation turns out to be a spavined old horse walking the rounds of an imaginary trapiche. Animals seem to get actually fond of such occupations. I remember an old billy-goat whose reluctance to furnish the motive-power of a baby-carriage had changed into such a passion for that employment that he would tolerate no rival on the track, and once killed a poor huckster's dog who, unintentionally enough, had excited his jealousy by drawing a larger-sized vehicle.

In process of time our four-footed ally may come to relish city odors, for his power of adaptation rivals that of the human species. In China, dogs eat rice; in Greenland, dried fish; in Siam, bananas; on the Pampas, carrion; and one of the Solomon Islands is inhabited by a race of half-wild curs that subsist entirely on crawfish. This plasticity of the canine species is almost enough to account for its infinite variety of forms: in the course of two or three thousand generations artificial selection may have turned a jackal into a mastiff, or a wolf into a pug-dog. It is strange to think what the continued operation of the same agency might have done for other animals, what marvels of beauty "in-and-in breeding," as our stock-raisers call it, would have developed from the gallinaceous tribes of the Old World, not to mention American parrots.

And what about the moral capabilities of such animals as monkeys and raccoons? Considering their intelligence, their faculty of imitation, and the mental superiority of a wild monkey to a wild dog, one cannot help thinking that the Darwinian theory might admit—not of an excuse, of course, but perhaps of practical demonstration.

CHAPTER IX

TRAPS.

MAN has been called a tool-making animal, and the first tool was probably a trap. I do not believe that our primogenitors were carnivorous. Long before they began to covet the flesh they probably hankered after the eggs and milk of their fellow-creatures, and had to devise means for catching them alive. They had no need of elaborate contrivances. Experience makes savages the best hunters, and it alone can explain their success in capturing animals whose cunning defies the best inventions of the amateur sportsman. With the simplest of all imaginable traps—an elastic stick with a noose—the Patagonian nomads catch hares, foxes, wolves, and the shyest of all American quadrupeds, the mountain-vicuña. Von Tschudi made the acquaintance of a Chilian farmer who had passed several years in the Andes before he succeeded in capturing a live vicuña. He had imitated the traps of the Indians, their method of fixing them in the sand of the river-banks, their precaution in obliterating the traces of their footsteps, but all in vain, till an Indian renegade revealed the secret,—namely, that the vicuñas invariably select their drinking-places where there is an audible ripple in the current of the river,—perhaps for the same reason that cows prefer a brook to a pond, and a running spring to a sluggish creek. The murmuring of the stream seemed to suggest the idea of purer and cooler water; and where the current was slow the Indians contrived to produce a ripple by an artificial obstruction.

Nearly every animal has some peculiarity or other that may be utilized for its capture. Minks have a queer passion for rummaging a pile of dry leaves, and the wild turkey can be taken in an open trap, because, for some reason, the idea of *going backward* never suggests itself to his mind. A Kentucky "turkey-pen" is simply a ditch with a roof of logs and ending in a *cul-de-sac*, but open at the other end. To this opening the turkeys are allured by "sprinklings" of corn or cranberries, and, entering the ditch where the bait is scattered more liberally, they follow it till they reach the *ne plus ultra* end; and it is a decided fact that such half-captives will poke around their pen for weeks without discovering the means of exit.

The female puma has a marvellous talent for hiding her lair, but the trapper knows enough if he discovers a place where she has torn her prey, for to that place she will return again and again, even after the carcass has been gnawed into a smooth skeleton. Jackals, too, are fond of revisiting the scenes of their former revels: some animals would seem to be endowed with the gift that supported Cardinal de Retz in his exile,—the faculty of "luxuriating on recollections." In Europe, where new preserves have often to be stocked with game-birds, hundreds of partridges are sometimes caught alive by the following simple device. Near the haunts of the game a brush-hedge with an opening here and there is set across a field, and on either side of the transit-holes the trapper fastens a wire noose. No bait is needed: partridges never fly over a hedge if they can crawl through, their motive being probably their general reluctance to betray their whereabouts by taking wing in an open field. Hunted conies, as well as rats and mice, are likewise almost sure to make for the next hole, incurring any risks for the sake of momentary concealment. In chasing a rat about a room, much trouble can be saved by twisting an old newspaper in the form of a sugar-loaf bag and placing it on the floor alongside of the wall. If the outlaw can be induced to approach it from the open side, he will dash in with a squeak of delight and can be captured before he discovers that his harbor of refuge has been blockaded.

Fear, however, is not the only motive of this mania for shelter-places. Of all animals, rodents are the most domestic. More than bees and swallows, and far more than man himself, they love to have "a roof overhead." They are fond of building a house within a house. The dryest corner of a dry-goods box in a snug old garret is the favorite dwelling-place of a house-rat, and even a church-mouse, the emblem of poverty, manages to find the wherewithal of feathering her little nest. It must be a poor place, indeed, where our sharp-toothed guests fail to make themselves comfortable. In the "Paradise of the Netherlands," the Elysian bog-meadows of Helvoetsluys, an acquaintance of mine built himself a model dairy with glazed brick walls and fire-brick foundations that seemed to defy the colonizing attempts of all rodents, but before the end of the first summer he found that all the same a pair of black rats had located and raised a family in his store-room. They must have clambered in through the window; but it puzzled him how they had managed their nest-hiding, till after a long search it appeared that they had excavated a forty-pound Dutch cheese, lined it with shreds of the tegumental canvas, and retired from the world, like friars into a fat convent.

The fish-otter burrows in the root-tangle of solitary river-banks, and hides the entrance so carefully that her nest can be discovered only by the sheerest accident; but she is often victimized by another foible which seems to be a peculiarity of the species. Otters are fond of *sliding*. In winter-time they scrape the snow from the top of a steep bank, and warm themselves whenever the sun comes out, but every now and then they fling themselves down, spread their legs, and shoot down-hill with all the delight of a school-boy trying a new sled. At the first sign of danger they disappear like a flash, for the end of their inclined plane communicates with a hole in the ice; but if you can watch them unobserved you may see them whisk up and down hill with evidently no other motive but fun. Their pastime, however, ceases to be funny if the hunter discovers their slide. After setting his trap near the foot of the slope and covering it with loose snow, he is almost sure of securing his game on the first sunny morning. Otters are also caught in traps smeared with oil of anise, for the smell of which they evince a queer passion. The olfactory predilections of different creatures are, indeed, almost as contradictory as their musical preferences. Chinamen confess that they can find neither system nor euphony in what we are pleased to call music; and beavers can be baited with assafœtida as readily as with castoreum. It is by no means impossible that our favorite perfumes, rose oil, cologne, and orange-water, may be downright torture to the noses of our domestic animals,—which would partly explain the ill humor of lap-dogs and similar pets that have to endure such luxuries all day long. Wild-cats, that disdain all other baits, can be trapped with valerian roots, and muskrats with stinkwort (*Anthemis fœtida*).

But it is a strange fact that such "medicines" serve their purpose only at their first introduction to a new district. After a certain number of muskrats have been victimized with stinkwort, the rest seem to arrive at the conclusion that the enjoyment of perfumes is one of the pleasures that kill. There is hardly any doubt that animals must possess some means of communicating their ideas. Arsenic has no perceptible taste or odor, and an ounce of it mixed with a bushel of cornmeal will destroy a cartload of sewer-rats in a single day; but all professional vermin-killers agree that such receipts lose their efficacy in a very short time. Somehow or other the survivors manage to trace the mischief to its cause; and old rats have been watched in the act of driving their young from a dish of poisoned hash. When the British first effected a settlement in Singapore, the traffic in monkeys soon became a regular branch of industry. The ubiquitous Chinamen used to go on trapping expeditions to the hills, at a time of the year when the mountain macaques were rather hard up for provisions and could be baited with "fuddle-cakes,"—*i.e.*, rice-bread soaked in a mixture of sugar and rum. The trapper used to hide behind a tree and let the monkey assemblage enjoy his bounty till their antics suggested that it was time for him to rush in, like Cyrus into the banquet-hall of Belshazzar. Experience, however, soon taught the little mountaineers to change their tactics. Instead of devouring the fuddle-cakes on the spot, they learned to gather them up and defer the feast till they reached a retreat where they could hope to be left alone in their glory. But the trappers, too, have since changed their plan. They manufacture a sort of narrow-necked jars, about the size of sarsaparilla-bottles, and, after filling them with a *mélange* of syrup and alcohol, they tie them firmly to the root of a tree and withdraw out of sight.

The monkeys come down and sip the nectar, a little at a time, till many a mickle has muddled their perceptives to the degree which the founder of Buddhism would have called the first stage of Nirvana,—indifference to

THE WAGES OF SIN.

earthly concernments in general. The trapper then approaches and collects his guests, whose exalted feelings often manifest themselves in a peculiar way. Some receive their captor with open arms, some hug their bottles with approbative grunts, while others lie on the ground, contemplating the sky in ecstatic silence.

In Mexico, monkeys are generally caught in box-traps. The Mexican farmer rarely kills a capuchin-monkey: man can afford to bear with his poor relatives where nature has been so kind to all, and in some districts of Oaxaca the *monitos* are as petulant as the sacred apes of Benares. Still, it is possible that this Hindoo-like forbearance of our next neighbors has something to do with their indolence, for I suspect that north of the Rio Grande the propensities of the long-fingered four-handers would "severely strain our tolerance," as Mr. Evarts said of the peculiar ethics of the Salt Lake Saints. Nor does the monkey-ridden ranchero object to their extermination by proxy: wherever maize is cultivated in the neighborhood of the river-forests the trapper is generally welcome. The box-trap method can be successfully employed only where the haunts of the game are well known, for the capuchins won't go out of their way without very special inducements, and in a field where monkeys have been caught before, their relatives become as circumspect as pickpockets in a metropolitan opera-house.

I once watched such a field for a whole afternoon before we caught one of the pilferers,—probably an outsider who had strolled in on the chance of getting a free lunch. The trapper had taken us to the loft of a corn-bin, where we could survey the garden and a portion of the adjacent woodlands; but our visitors gave the *trampa* a wide berth, though their sidelong glances showed that they had not forgotten the place. They nosed about in the stubble, but the gleaners had been beforehand with them, and, after an apparently aimless ramble among the furrows, one gray-headed old brigand happened to fetch up right in front of the box, perhaps much to his own surprise, though, by a strange coincidence, five of his relatives reached there at about the same time. The gray-head raised himself on his hind-legs, surveyed the field in every direction, and then began to *tack*,—*i.e.*, approached the box in a roundabout way, lowering his head with every step, till he reached the trap-hole, where he settled down on his haunches, but with one hand in position for a sudden back-spring. With the other he explored the sand for maize-kernels, —*in utrumque paratus*. His manœuvres seemed to have reacted on the imagination of his companions, for two youngsters took to their heels, but presently returned, and, with a bravado-flourish of their tails, advanced to within a few inches of the *trampa*. Turning half round, to let the by-standers witness the proofs of his heroism, one of them stretched out his hand till it all but touched the suspicious apparatus, but then, as if struck with a sudden horror at the thought of possible consequences, they both fled more precipitately than before and rushed into the arms of a motherly old capuchina, whose caresses at last composed their overwrought feelings.

Their comrades in the mean time encroached upon the old man's claim till he got up to reconnoitre the rear of the *trampa*. Not a crumb was in sight, but his sense of smell seemed to reveal the riches within, for, after a careful examination of the bottom-board, he raised his head with an impatient jerk and a glance in the direction of the farm-house: "A sad comment on modern hospitality and Christian civilization!"

The party at the other end crowded around the entrance and grinned at each other with excitement. The *trampa* had been baited with *meladas*,—ripe corn-ears boiled in treacle: the corn alone would have been enticing enough, but the syrup made it so nearly irresistible that every now and then one of the tempted capuchins had to ease his nerves by a rush to the rear or a spring against the picket-fence. The return of the sachem saved them from greater indiscretions, for some of them had already begun to measure the depth of the box with their outstretched arms. Before long, however, their impatience revived: the junior members of the conference seemed to advocate an attack *en masse*, but before matters came to a crisis a new-comer appeared on the stage,—a lank and hungry-looking stranger, who seemed to have emerged from an elder-coppice behind the fence. For a minute or two he watched the proceedings in silence, and then marched straight upon the

TRAPS. 233

box, evidently resolved to astonish the natives. He

"IN FOR IT."

squeezed himself through the trap-hole with the success

that often attends a bold enterprise, and easily obtained possession of ten select *meladas*,—a monopoly, in fact, for in the next moment his pre-emption was ratified by the fall of the tail-board. The sachem turned round with a coughing grunt,—"I told you so,"—then took to his heels, and, seized with a sudden panic, the whole troop fled, shrieking, in the direction of the high timber.

Monkeys caught in steel-traps are wildly obstreperous and can hardly ever be tamed: their first experience of man's inhumanity to man seems to have left an indelible impression. Steel-traps make their captives ferocious. There is something in the arrangement and *modus operandi* of the treacherous implement that appears to aggravate the horrors of the result and excite the wrath of a naturally savage animal to the raging pitch: the struggles of a captured wild-cat sometimes liberate her at the cost of a limb, and, if the iron has not collared her at the very middle of the neck, she will tear herself out at the risk of leaving her scalp behind.

Pitfalls have the opposite effect: they cow their prisoners; the darkness and mystery of the predicament and the uncertainty of the result seem to paralyze their energies. In Abyssinia, where all our principal menagerie-men have an agent or two, sand-foxes, jackals, and even hyenas are often caught in pits and taken alive by a very simple process: the hunter goes down, lariats his quailing captives, and, while his partner draws the rope tight, he ties their four legs, gags them with a thong of raw-hide, and the dire *feræ* are dragged out like butcher-calves.

Practical naturalists are generally the most successful

DECOY OWLS.

trappers, for Lord Bacon is probably right, that observation is quite as prolific a mother of inventions as necessity.

Only observation could have revealed the fact that little song-birds can be attracted by the sight of a bird of prey. A common chicken-hawk will serve that purpose. Fasten a tame hawk to a bush, and before the end of an hour all the finches and thrushes of the township will find it out and meet in general convention,—an indignation-meeting, perhaps; though it is hard to understand what they can hope to accomplish against an enemy who could kill a score of them in ten minutes. *A priori*, their proceedings would seem as incredible as an assembly of mice around a chained cat. But the experiment never fails: a hawk, an eagle, but especially a ferocious-looking old horn-owl, will allure birds at a time when they would disdain to neglect their domestic business for the sake of any tidbit. An owl-riot they seem to consider as a sort of public duty which must take precedence of all other affairs, for even migratory birds will stoop from their flight through air and light to screech around an old night-spectre. In Northern Italy, where game is scarce, every farmer has a tame *buba* and a potful of birdlime, and thousands of Northern songsters, hastening fondly home from their winter-quarters on the Mediterranean, fall a victim to their ruling passion and perish in exile,— "butchered to make a Roman holiday."

CHAPTER X.

FOUR-FOOTED PRIZE-FIGHTERS.

In Anglo-Saxondom circus-combats have gone out of fashion. The efforts of Bergh & Co. have promoted the introduction of less destructive, if not more instructive, amusements, though, as Herbert Spencer observes, all our more exciting pastimes are still prize-fights in disguise. But in the lands of the Latin races the undisguised form of the sport is still too popular to be illegal, and frequent enough to enable even unwilling spectators to convince themselves of one curious fact,—viz., that death in the arena must, on the whole, have been the least disagreeable way of crossing the Styx. It is the easiest death. The old Berserkers knew what they were about when they prayed to die in battle rather than in bed: in the heat of combat wounds are actually unfelt; excitement operates like an anæsthetic, and the fighter reels into Nirvana as in a trance. A rough-and-tumble fight is far more exciting than the machine-war of our modern armies, but even modern soldiers know that, in battle, injuries not involving the demolishment of a motive organ often remain unnoticed till they announce themselves through exhaustion or such external symptoms as swelling and hemorrhage. After the repulse of Torres Vedras, Massena congratulated the survivors of his staff and vaunted himself bullet-proof, when the remark of a by-stander caused him to put his hand to his wounded forehead: "*Chien de Notre Dame! qu'est-ce que ça?*" said he with unfeigned surprise. Count Ranzau, the *Streithans*, "Rowdy Jack," as his comrades called him, once received three stabs before he knew that he was hurt; and in the battle of Ostrolenka, Kosciusko led his "scythe-brigade" till his horse was shot down, when, dismounting, he found himself crippled by a shot that had struck him an hour before,—merely through the boot-leg, as he had thought at the time.

Not all soldiers are volunteers, and cowards, as Shakespeare says, "die many times;" but a circus-manager would have no difficulty in raising a regiment of bullies, from a count to a cock-bantam, not only willing but impatiently eager to try conclusions, with or without a referee. Marcus Aurelius provoked a fierce revolt by trying to compel the gladiators to fight with blunted swords (Xiphilin., l. xxi., 29), and four-footed champions with a rival in sight often fall upon the biped who tries to restrain them. Warfare is the normal medium of natural selection, and captive wild animals, of the carnivorous species particularly, need very little encouragement to accept a challenge.

An instinctive recognition of these facts, rather than of our ethical objections, seems to prevent semi-civilized nations from seeing anything wrong in a prize-fight. "*Volenti non fit injuria.*" They might think it disgraceful to plague a peaceful creature, but can see nothing objectionable in witnessing a display of natural combativeness. "*Que idea!*" exclaimed a Mexican whom Bishop Riley had taken to task for his cock-fighting proclivities. "*Que daño hay?* They volunteer performances on every dung-hill : are they any the worse for having spectators ?" The historian of "European Morals" (vol. i. p. 290) observes that in Spain an intense passion for bull-fights is quite compatible with a charitable disposition; and the Hindoos, with all their Buddhistic prejudices, are enthusiastic votaries of the cockpit. Beast-fights were the most popular amusements among the ancients. King Porus of India, who was probably either a Buddhist or a Brahman, entertained his conqueror with what the Spaniards would call a *gran matanza* of trained elephants. Nebuchadnezzar had his famous lion-pit ; Prusias, the King of Bithynia, imported Indian tigers ; and Antiochus Epiphanes kept a lot of fighting-bulls. But these private sports were dwarfed by the public *circenses* of Imperial Rome. Three hundred bears were let loose during the games of Claudius, three hundred lions and five hundred bears at the triumph of Hadrian, and at the dedication of the Coliseum by Titus *five thousand* wild animals on a single day! (Magnin, " Origines du Théâtre," pp. 449-453.) Tigers, elephants, rhinoceroses, hippopotami, giraffes, and lions were imported in numbers that must have employed an army of hunters and trappers. The Numidian satraps were *ex-officio* agents for the Roman menagerie-dépôts (Friedländer, pp. 141–145), and the African proconsuls were specially instructed to keep a lookout for novelties, "*quidquid novum ex Africa:*" the big snake of Bagradas or an able-bodied unicorn would at once have made the fortune of its captors. Pliny's "Natural History" abounds with arena statistics, mingled with curious anecdotes and still stranger superstitions, though certainly no other zoologist had ever such opportunities for studying the nature and habits of wild animals.

During the Middle Ages the Spanish Moriscoes were the best naturalists. Their intercourse with the Eastern Caliphate filled their cities with outlandish curiosities, and some of the princes of Cordova were great sportsmen: Abu Abdallah and Abdel Zagal used to import African lions and bait them with a special breed of mastiffs. Their Christian successors seem to have inherited that passion, and when the African *feræ* became scarce they found a good substitute in the half-wild bulls of the pastoral Sierras. The Andalusian *toros bravos* were at first baited with dogs, but the kings of Aragon introduced trained swordsmen, and bull-fighting then became a national passion. Saragossa, Malaga, and Madrid vied in the splendor of their matanzas, and at the end of the fifteenth century all the towns and larger villages, and even the wealthier convents, had their special bull-rings.

Four successive popes tried in vain to stop the game. Some of them threatened excommunication; but they found that their *bulls* did not scare the toreros, and Gregory XIII. had actually to revoke his own edict: nay, the clamors of the Spanish clergy obliged Clement VII. to pass a special ordinance legalizing bull-fights on church festivals! (Lecky's "History of Rationalism," vol. i. p. 308.) In the cities the matanzas went on as merrily as ever: Seville had a special school for toreros, and Philip the Second kept a torero guard and a chief court matador. Three hundred years of monk-rule and misfortune have not tamed this passion. Cadiz, Cordova, Toledo, Medellin, Cartagena, and Alicante—mere beggar-towns, compared with their former splendor—still manage to get up a weekly matanza. No saint can hope to rival the popularity of a successful matador: the French publisher Hallerman made a fortune by chromotyping the portrait of the torero Perez. José Maria Perez began his career as a Cartagena *canallon*, or circus-sweeper, and, in spite of his dissolute habits, died the richest man of his native town. His arrival in a bull-ring city produced a regular *furore*: merchants closed their offices and teachers their schools,* disguised monks

---

* *Natura si furca expellas,* etc. Last August [1881] a Georgia moon-shiner captured a wild-cat and brought it to Birmingham, Alabama. The dignitaries of that city assembled at the court-house and resolved by ac-clamation,—1st, to pit the cat against a certain town-dog; 2d, to celebrate the event by a general holiday. On the following morning all work was

escaped from the convents and mingled with the lowest rabble to enter the arena unperceived, paupers pawned their last coat to raise the requisite *real*. A similar enthusiasm often gathers round a victorious bull. The chief advantage of the torero is not the clumsiness or the good nature but the stupidity of the average *toro*. A Catalan bull can dodge and turn like a cat, but, for all that, can be taken in by tricks that would not fool a pig. Practice, however, makes him a ticklish customer, and a bull who has killed his man is in a fair way to become himself a matador. "If bulls could be trained," says the naturalist Azara, "they could be made as dangerous as a horseman armed with a pitchfork. But some toros contrive to train themselves, and the public love of excitement is then gratified with a vengeance. In 1835, in the midst of the civil war, a Barcelona bull became a municipal idol, the object of a regular Apis-worship. When he had killed five men and ten or twelve horses, the yard around his stable was thronged with devotees, though his keeper, fearing foul play, would admit no stranger to the interior of the sanctuary. After his last victory on the festival of San Antonio, the crowd went almost crazy with excitement, under deafening cheers and a continual shout of "*Bollos por el toro!*"—"Cakes for the bull;" a libation of reals came down like a shower,

---

suspended, and the population of Birmingham formed a ring, while the mayor held the stakes. The cat won in two rounds." (*New York Weekly Herald.*)

FOUR-FOOTED PRIZE-FIGHTERS. 243

and when the victor was dragged out of the gate, a young girl, who had got a prize in a pantomime, leaned over the balcony and, at the risk of being impaled, crowned the gory brute with her own garland. This Apis was at last vanquished by the Aragon matador Zorilla, who boasted that he had never invoked the aid

A RELUCTANT TRIUMPHATOR.

of the *chulos* nor permitted any beast to reduce him to defensive shifts.

Such matadors might change their tactics in Burmah, where wild elephants are pitted against horsemen and *cornacs* (elephant-riders) and often rout their trained relatives by the reckless fury of their attacks. A cornac's office is no sinecure: he has to stick to his seat while his *hutti* squirts and careers around like an exploding locomotive or encounters his adversary with the force of a catapult. Fighting elephants guard their trunks by doubling them up like a clinched fist, while using their heads like battering-rams, or they stand shoulder to shoulder, after the manner of fighting boars, and, after a prelude of sidelong pushes, suddenly hew away at each other with their tusks. There is not much danger of a general breakdown, for the legs of a full-grown elephant will sustain him in a collision that would ditch a four-horse team; but there are greater perils: the wild elephant may get the upper grip and pull the rider from his seat, or the trained hutti may "get mad." No elephant can be entirely trusted: the tamest of them are subject to tantrums, often most malapropos. During the progress of the duel the hutti seems to forget or ignore his rider; but if he has received a fatal wound the cornacs have to jump off and run for their lives, experience having shown that wounded elephants generally expire in a paroxysm of rage. The feeling of approaching death seems to inspire them with a sudden fury against the authors of their misfortune. A similar outbreak of savagery *in articulo mortis* has been observed in other animals: chacma baboons and tame panthers in their last hour often drop the mask of allegiance, like Lucius Vanini, "determined to die free."

Since the abolition of their cruel religious ceremonies, beast-fights seem to form the chief pastimes of the Hindostan princes. The largest walled circus of modern times is in Baroda, where the Guicowar has a special park with elephants, panthers, and rhinoceroses enough to get up a bi-weekly fight,—and no sham fight, either (Louis Rousselet's "India," chap. vi.). He has a troop of drilled matadors,—" elephantadors," as Rousselet calls them,—besides trainers and hunters, and has paid as much as eight hundred dollars for a good hutti.

Domesticated elephants, however, have to be fuddled with bangh to excite their combativeness, and their training is so expensive that rajahs of moderate means prefer prize-fighters *per naturam*,—panthers and wild boars. Ranjit Sing, the Maharajah of Dholepore, used to keep a park of picked tigers that were fed on live dogs and pitted against all the wild beasts his hunters could lay hands on. One of these tigers, an enormous brute with a head like an ogre, was presented to General Havelock, and thus found its way to Lucknow, but it was finally sent back to the maharajah's successor, who had set his heart on having the best fighting-tigers in India. The last Nizam of Hyderabad had a tame cheetah that followed him in all his campaigns and enjoyed all the privileges of a court favorite; nay, Aga Muhamed, the Guicowar of Guzerat, kept a carnivorous horse, an unnatural brute, which once, in the presence of Professor Schlagintweit, knocked down a goat and devoured its udder before he could be driven off. Only his old acquaintances could manage him; strangers he was very apt to assail with his teeth, and the dogs which were sacrificed to his appetite he tore to pieces with the energy of a ravenous wolf. But a still more famous fighter was Black Jan, the pet of the Rajah of Samarang in Java. Jan was a Sunda panther, born in captivity, whose constant practice in the arena had endowed him with the nimbleness and bloodthirst of a ferret. His matchless skill in defending himself against adversaries of superior size often attracted the planters of the neighboring Dutch settlements, and even visitors from Batavia. A Batavia journal describes him as rather under-sized for an adult specimen of the *Pardus javanensis*, but remarkably stout-limbed, and "not agile but agility itself." Experience had acquainted him with the weak points of all possible antagonists. A wild boar he demolished by leaping upon his back and belaboring his head with his claws, a bull by fastening his teeth in his throat. Dogs he fought in the regular cat-fashion,—by striking at their eyes and collaring them at the first opportunity. If a pack of them tackled him at the same time, he would retreat to a corner and keep them at bay till he saw a chance for a head-spring, his favorite trick on an enemy with a dislocable neck. Jan was the idol of Samarang; but the peasants of the neighborhood suspected his owner of witchcraft, and in the circus the restive visitors often broke out in groans when the "pet" made his appearance.

During the Middle Ages the nations of Europe vied in bull- and bear-fights and badger-baitings; but those times are past, and only on the lower Danube can such *circenses* still be carried on in public. In Eastern Hungary, where landed proprietors are permitted to select recruits and appoint their own tax-collectors, feudalism

THE RAJAH'S PET.

has still vitality enough to make every nobleman the king of his domain, and even in the neighborhood of Buda-Pesth the government does not interfere with national pastimes as long as the magyars will forbear the more dangerous game of national politics. A two-horse *kutsche* with a trio of cavaliers crossing the Buda bridge of a Sunday morning generally means that there is *etwas los*,—something up,—some fun ahead on one of the up-river country-seats. The Carpathian Mountains still abound with wolves, and in the Bakony Wald wild-cats and wild boars are caught every week in the year. *Konök-Deresch*, or wolf-baiting, is a sport which has perhaps been imported from the region where the ancestors of the Magyars hunted the jackals of Imaus. They use a sort of lariat of untwisted strands of hemp or horse-hair that sink between the teeth of an animal trying to gnaw it. With a rope of that sort the wolf is fastened to a picket-stake by means of a ring that permits him to run round and round without entangling his tether. On the Danube curs are cheap, and if the wolf proves a good fighter he may hope to live and fight another day, or even to advance to the rank of a household pet. If he turns tail, his fate overtakes him at the end of his tether, and his carcass is used to instruct young shepherd dogs in the higher branches of their profession. Near Pesth, where wolves are rare, the wild boars of the Bakony Wald act the leading *rôle* in the game of *Deresch*. On account of the peculiar formation of his neck, Lord Bacon cannot be tethered, so they turn him loose in a corral with an amphitheatre of hay-bales and reserved seats on the wood-pile. An old boar is by far a more dangerous customer than a wolf. A well-aimed cut of his knife-like tusks will rip a dog from neck to stern; but trained hounds checkmate that game by the "catch and vault" trick,—*i.e.*, they grab the tusker's ear and jump over his back, and thus keep his head in chancery till hunters or comrades come to his assistance. Even in Hungary a good boar-hound is worth ten florins; but the Magyars are a magnanimous race, and if the boar contrives to vindicate the dignity of nature they are apt to reward him by an unconditional pardon,—*i.e.*, to open the gate and let him depart in peace.

Now and then one of the Nagy Tassar, or "Big Squires," manages to get hold of a bear; and during my last visit in Buda Dr. S—— took me out to the Raitzen suburb, where a champion of that sort was on exhibition. We found a big, fat he-bear, whose owner had sent him down from Komorn, but, having neither a fit locality nor an opponent worthy of his prowess, they were not going to fight him, but merely to exhibit his fencing skill. The pest of Pesth is the brown rat. Legions of rats infest the 'longshore quarters, and every *Roskam* (livery-stable) has ratters for sale,—a sort of shaggy pinchers,—as aggressive as any bull-dog. Of these curs our host had a whole brigade, and nearly every one of his visitors had brought a recruit or two. We picked out eight of the pluckiest and ugliest,—worth about forty cents apiece. Before we let them loose they had smelt the bear and scratched savagely at the door; but their simultaneous appearance did not disconcert Bruin in the least. He was taking his breakfast in the corner of an empty chamber, and when the door opened he did not even interrupt his meal, but with the utmost good humor flung the puppies against the wall as fast as they came on. They charged him again and again, but they did not "rile" him a bit: whenever we called his name he looked up with his mouth full of corn-cake or responded with a complacent grunt, while he attended to the dogs in a sort of absent-minded way as a man would to a swarm of flies. But, with all his nonchalance, he knew exactly what he was about: nearly every slap was a hit, and every hit did the business for that particular pincher. When four of them lay howling and grovelling in the opposite corner of the room, the rest became meditative and waited for special instructions before they renewed the combat. Like the victims of the Minotaur, they bayed him from a distance, jumping left and right, with an occasional advance whenever he licked the bottom of his breakfast-pail, for, though he could have routed them by a mere gesture, he did not think it worth his while. He sat down and began to lick his paws, till we were going to leave the room, when he got up and followed us to the door. The moment he turned his back the dogs made a dash, and one of them nabbed him from behind, but in the same instant, almost, he went spinning through the air and with a crash against the board of the opposite wall. The bear had turned like a shot and struck his assailant before a man could have lifted a stick. No boxer could have parried the electric suddenness of that blow, which was nevertheless delivered with the force of a sledge-hammer stroke, for the cur was at least a twelve-pounder, and his collision with the wall actually made the windows rattle. We whistled off three of the dogs, while their disabled comrades were left alone with their conqueror. But his was evidently not a rancorous soul: when we opened the door half an hour after, he was sitting near the window licking one of the curs as a bitch would a lame puppy.

Even our big grizzly does not deserve his ferine reputation. A fellow combining the strength of an urochs with the claws of an ant-bear is naturally not disposed to put up with insults, but his habits in captivity prove that he prefers sweetmeats to flesh; and, though in stress of circumstances he stills his hunger without fear or ceremony, he never indulges in the wanton destructiveness of the panther. In the summer of 1879 a grizzly bear entered the enclosure of Alexander Presswood's farm, near Jacksboro', Texas, and helped himself to a quarter of venison that was hanging on the shady side of the farm-house. Near the back door a little boy had fallen asleep on a pile of wool, while his still younger sister was playing at his feet. Seeing the child move, the bear came up and examined it, and then sniffed around the head of the sleeping boy, who, suddenly awakening, started up and slapped the grizzly in the face. The bear retreated and trotted off toward the fence, closely followed by the boy, and, to the horror of his mother, whose attention had been attracted by his angry exclamations, the little fellow raised his foot and dismissed the brute with a farewell kick as he squeezed himself through the narrow gate.

The most truculent of all carnivorous animals is perhaps the little pine-marten or martern (*Mustela martes*), a creature about the size of a fox-squirrel, but capable of killing ten times his own weight in poultry before a squirrel could eat a nut. If one of them gets into a pigeon-house he is apt to make a night of it; *i.e.*, he will butcher away till daylight interrupts him. Charles Sealsfield, who built himself a chalet near Brunnen, in Switzerland, once caught a pine-marten *in flagranti*, and, on inspecting the loft of his poultry-house, found forty dead turkeys and half a hundred chickens and pigeons. The murderer had contented himself with tearing their throats: some of the short-necked hens showed no visible injury, and all were in what a poulterer would call a marketable condition. Such wholesale destructiveness can sometimes be explained by the needs of a burrowful of hungry whelps; but pine-martens leave their victims where they drop; the female suckles her kittens till they can shift for themselves, and never brings any meat home. The little wretches can be trained to fight, and will attack kids, hares, and even pigs; but, with all their bloodthirst, they are arrant cowards whenever they meet a less helpless creature: the mere sight of a dog is enough to scare them into a mouse-hole. In March the males fight with such a craziness of rage that they tumble from the trees and roll around in the grass, where they have sometimes been killed with a common cudgel. Sealsfield describes a combat of such duellists. They chased each other round and round a tree, through hollow roots and bushes, squeaking, hissing, and barking, and every now and then clapperclawing and snapping away like little wild-cats. His repeated intervention merely caused them to confine their scuffles to the higher branches, but after each round they raced up and down the tree and often whisked by within two or three yards of his feet. Fighting-cocks are even more tenacious, and the Alpine ruff, or rock-plover (*Tringa pugnax*), is often captured during the progress of his desperate monomachies. But birds lack the vindictiveness of four-footed prize-fighters. If a fighting-cock gets killed, it is mostly on account of his own obstinacy in preferring death to the alternative of saving himself by flight; but a marten has to fight it out willy-nilly,—the victor generally kills his rival. Besides answering the purposes of natural selection, such honeymoon combats may serve to check the increase of noxious creatures that have no natural enemies to pay them in their own coin. Being semi-nocturnal, and excellent runners, swimmers, diggers, leapers, and climbers, martens are very hard to exterminate, and would become a worse nuisance than rats if the progenitors of the species did not attend to each other. Their relatives the European ferrets are the implacable foes of the whole rodent species. The formation of a ferret's body is wonderfully adapted to facilitate its special business. It attains a length of two feet and a weight of five or six pounds, but the whole body can be drawn through a napkin-ring. The legs are very short and remarkably far apart; the occiput tapers toward the neck, and the rump toward the tail; a full-grown ferret can squeeze its head through a rat-hole, and where the head goes the rest of the body follows like a caudal appendage. No other mammal bears such a striking resemblance to a snake. In proportion to his size, an old ferret is an amazing tough customer, and can be trained to clean out a whole rabbit-colony and drag the settlers out of their holes; and that seems, after all, his proper vocation, for in the rat-business he is rather liable to "get stuck,"—*i.e.*, to squeeze himself into a hole with a tight place where he can neither advance nor retreat, and thus risks falling a prey to his intended victims, which are not slow to take advantage of his "fix."

Like bears, dogs are by nature far less savage than the *felides*, and yet it is from the canine species that artificial selection has evolved the ultra-type of reckless ferocity. The boldness of a bull-dog is different from that of any other wild beast: *courage* is not the word to describe his disposition: he is not satisfied with defending himself or his master, he is not stubbornly valiant merely, but blindly aggressive, combative from a sheer love of combat, without the least regard to the merits of the cause or the advantages of the result. The mere sight of a stranger—biped or quadruped—is enough to throw him into a fit of that fury which *hashish* is said to produce in the human animal; he is in a chronic state of *furor litis*, ready to run amuck at the first opportunity. Under a real provocation this truculence rises to a perfect frenzy: in his efforts to break his tether, an angry bull-terrier will tear the hide of his neck into shreds or snap his teeth on an iron chain, and, if he can break loose, danger will count for nothing against the rage of glutting his revenge. The prospect of certain death may be said to have no terrors for a thorough-bred fighting-dog. Spanish wolf-dogs will successively rush upon a bear whose paw has smashed every comer at the first blow. A Danish mastiff will go headlong upon a man with a levelled shot-gun. Nay, Baron Gaisner, a well-known Vienna sportsman, laid a wager that his rat-terrier would tackle a big bloodhound; and at the word of command the little dog won the bet by losing his life.

Some farm-dogs do not even wait for commands to fly at every stranger passing their premises. Three years ago a large panther escaped from a menagerie-man who had pitched his tents near Lansing, Michigan. Toward evening the deserter appeared at the door of a wayside smithy some three miles south of Lansing. The smith flung a piece of coal at his head, and the panther trotted off, and was passing the fence of an orchard, when a vicious-looking cur leaped over the enclosure and without a moment's hesitation fell upon the refugee, who was peacefully jogging along toward Ann Arbor. "Three seconds later," says the Detroit *Press*, "any liberal man would have given five dollars to know what that dog thought of himself."

Old fighters, however, generally know what they have to expect, and go it headlong,—

> Den Bären gleich, die keine Wunde scheuen,—

taking and giving wounds with equal recklessness. There are animals of such thick-headed stolidity that their fortitude needs not much stoicism; but next to a monkey a dog is nearly the most sensitive of all vertebrate creatures, and his power of endurance under certain circumstances can be explained only by the anæsthetic influence of excitement. Maimed, blinded, and disembowelled, a boar-hound will yet stick to his foe with the tenacity of a snapping-turtle, and an English bull-dog will fight while he can stir, resolved to yield only in yielding his life.

Dog-fights are represented on the bas-reliefs of Persepolis, and formed probably the earliest pastime of the pastoral Aryans. *Hund* (hound) was a favorite cognomen of the ancient Germans, who prized valor as the supreme virtue; the four-footed fighter *par excellence* became the companion of the biped warrior, and only among the Semitic nations the aversion to the uncleanliness of man's truest friend outweighed this partiality. The Saracens shared that prejudice; on the treeless plains of their native country, where every herder is a horseman and hunters can see their game from afar, dogs are, indeed, less indispensable; but the Spaniards valued a stanch dog above a fleet horse, and were the first to breed those big blood-hounds that proved their terrible efficacy in the conquest of the New World. The race of the Caribs that inhabited the east coast of Central America and the larger islands of the West-Indian Archipelago was almost exterminated by these domestic beasts of prey. Davila Pedrarias invaded Panama with three hundred troopers and forty mastiffs that had been trained to fight in ranks and used to charge in the van of the squadron; and Navarete quotes as the lowest estimate that, in Cuba alone, the blood-hounds of Victor Holgar killed four thousand natives in a single year! Balboa's famous "Adjutant," Leon-icico, was a gigantic butcher-dog that could kill an Indian as a terrier would despatch a rat. This monster wore a coat of mail, and, in the opinion of his master, was worth any ten cuirassiers in the Spanish army, for in the three campaigns against the Honduras hill-tribes he had rid the "king's dominions" of more than two hundred rebels. During the last year of his eventful career he drew the pay of a color-sergeant, and used to be carried on horseback to economize his valuable strength. The Indians hated him like a were-wolf, and their cazique had offered a large prize for his head, but that cursed cuirass always saved his life till a well-aimed arrow hit him in the eye; and if he went to where he belonged, his brother Cerberus could apply for a furlough.

The "Aragon hounds" of Northern Mexico are supposed to be the descendants of this breed. Their wild life in the Sierra has added something wolfish and outlandish to the savageness of their appearance, but they lack the stubborn courage of their ancestors, and I have seen one of them beaten by a common tramp-dog. Among the Mexican sportsmen the excitement of a dog-fight is enhanced by a subjective interest. They all bet. Bets, moreover, have to be paid on the spot, and the backers of a losing brute often revenge themselves after the manner of true savages, though they would probably call it the old Roman fashion. In a Puebla museum I was once looking at a panorama of the famous circus-scene, where the spectators, *pollice verso*, are clamoring for the death of a fallen gladiator; but a Mexican caballero, after listening to the comments of my companion, suggested that these clamors might be justified by *the disappointment of the heavy betters*,—for the prostrate hero looked really twice as large as his victor. Two months after, I realized the meaning of the caballero's remark. The burghers of Medellin had got up a *gran funcion* between a young bear and a butcher-dog. The bear, being more than half grown, was largely the favorite, but, after an obstreperous scuffle of ten or twelve minutes, skill prevailed over brutal strength, and the backers of the vanquished plantigrade avenged their loss by giving him a terrible beating.

Dogs do not eat their conquered foes, as bears, and

"VÆ VICTIS."

even boars, are apt to do; but it is a curious fact that they fight best after a long fast. It *whets* their mettle, as sportsmen express it. Frederick the Great found one consolation in the vandalism of the Russian invaders,—it exasperated his men; and a *bona fide* fast seems to produce a similar effect. In the Rhamadan season strict Moslems eat only every other day, and Burckhardt advises strangers to approach them on those *other* days: starving, instead of improving their temper, puts them into an aggressive mood. The famished anchorites of the Nitrian Desert were dreaded like so many wild beasts; "maceration," as they call it, may have answered its purpose in subduing some other propensities, but it certainly excited their combativeness; and I have often wondered if it would not be a good plan for a commanding officer on the eve of a battle to order a general fast-day, with a promise of double rations after the *Te Deum*. The well-fed Medes were beaten by the starved Persians, six Roman generals by Spartacus with his hungry outlaws, the Visigoths by the Saracens, the Austrians by the Sans-Culottes. The heroes of the Crimea were perhaps too outrageously starved, but the feat of Balaklava would hardly have been achieved by a *full* brigade; and I cannot help thinking that even the efficiency of our Dixie mamelukes had something to do with the deficiencies of their commissariat.

In North America, too, "dog-fights in a ring" are still very popular, and more frequent than Mr. Bergh may imagine. But the most passionate devotees of the sport are the burghers of the Dutch seaport towns. "A sad comment," etc.; but, as Mr. Bruce's boy remarked, "People wants to have some fun." North Holland is getting rather barren of out-door sports; in a land of truck-farms fox-hunts are out of the question, wild ducks are getting scarce, and every game-preserve is watched like a young ladies' seminary. And, besides, though the Hollanders have ceased to be a conservative nation, many of their by-laws still date from a time when prize-fights were patronized by princes and priests, and the Amsterdam jonkers need not go very far out of town to indulge in things which in England could be explained only by the sheriff's "connivance with both eyes."

"*Sog, wo sal hij stryten?*" ("Where is he going to fight, I wonder?") is a frequent remark on meeting a fair specimen of the *gryffhond*, a sort of mastiff,—nobody doubting that the *hond* is kept for fighting purposes. A rendezvous in Muidenhaven means generally an invitation to a dog-fight. Northeast of the main harbor extends a long line of private wharves, flanked with promenades and villas and here and there with public restaurants. A special variety of these restaurants is the *gardenhuys*,—a tavern licensed to dispense refreshments, but without a sign-board, and therefore safe against the intrusion of unintroduced strangers,—a sort of club-house, with a *factotum* president. The proprietor of a gardenhuys generally keeps a ten-pin alley, often a cock-pit, and sometimes a pigeon-shooting-gallery, but nearly always a dog-ring. He keeps fighting-dogs of all kinds, gryffhonds, terriers, and pinchers, but permits his guests to make his ring the arena of their private honds. Some of these fighting-dogs have achieved a national reputation. The competition for the puppies of a favorite gryffhond rivals the wrangle over the bulbs of the famous tulips of old, and the professional fanciers keep regular blue-books of dog-pedigrees. A fighting-dog does not lose caste by being overpowered in one or two rounds: only death, a permanently disabling wound, and the refusal to "come to the scratch," constitute an absolute defeat. Even a defeated hond, though his rank is lost, may recover a *quasi* prestige by killing his adversary in the next fight; but there are dog-dynasties that have preserved a clean record for five or six generations; and in Amsterdam my brother once procured me an introduction to the most invariably triumphant warrior of his age,—Klaas, the *Koning*, a mastiff of doubtful descent, but of a most indubitable superiority over all his living rivals. His owner, a choleric old skipper, had inherited him from a relative who took no interest in pedigrees, but the Koning's victories had founded a new peerage, and his descendants began to eclipse the *ci-devant* aristocracy of the neighboring towns. The "King" deserved his rank. He had never lost a fight. His owner had pitted him against boars, bulls, and several of the outlandish brutes which the Dutch colonies inflict on the mother-country, but he had never failed either to kill or to rout his foe. His triumphs became such foregone conclusions that the bets were chiefly *against time*,—wagers on his ability to crush his foe in more or less than so many minutes. In 1875, Klaas had been king for three years, and his courtiers became so numerous that his master got tired of their visits and sent him every Sunday to an inn on the Prinzengraacht, where he received callers from nine to eleven A.M. No pasha of nine tails could have displayed more conscious dignity. At home Klaas had the reputation of being the laziest dog in North Holland, but in the hotel he declined to sit down. He seemed to know that the guests had come for his sake, and kept walking up and down with a leonine strut, now and then vouchsafing to accept the homage of a new visitor or to acknowledge the greeting of an old acquaintance. Strange dogs he received with a stiff *grandezza*. He refused to permit them any familiarities, but sometimes scrutinized the big ones with a sort of professional interest. They took care to give him a wide berth. Klaas weighed two hundred pounds, but there was not an ounce of superfluous tissue under his hide, unless a number of welt-like scars could be considered expletive. Toy-terriers, though, will rush in where not angels only but bull-dogs would fear to tread, and there were cases on record of several puny yelpers having done their utmost to provoke the King's wrath.

On such occasions his majesty would pretend to be asleep; but if his assailant insisted on waking him, he would look up, not at the cur, but at the cur's master: "Couldn't you save me the necessity of demeaning myself?"

Down-town he had sometimes been attacked by a junta of street-dogs, but it wasn't quite easy to scare him. When he crouched for a spring there was something in his look that rarely failed to make the front ranks unpopular, and the allies generally retreated in time to save their vertebræ. "I wouldn't mind pitting him against any two dogs in Holland," his master told me, "but there is one thing I am afraid of: he has a weak spot, a bad scar under his left jaw, and by the way he fights I see that he knows it. Against one dog he can hold his own in spite of that, but two—if one of them should manage to collar him from the left, I do not know what mightn't happen. There is a dog in Groningen, they say, can beat him," he added in a confidential whisper, "a butcher-dog from Helderdam, but, unless he is the devil himself, I guess Klaas knows a trick or two that will stop their bragging."

At home the Koning passed the larger part of the day behind the *Kacheloven*, the great brick stove that still warms the dwelling of the orthodox Hollander, and burglars could have abstracted the rugs from under his very nose. Klaas never interfered in domestic affairs, and even disdained to beg for soup-bones: he knew they could not afford to starve him. But on the day of battle he was a changed dog. They used to take him out in an open cart, and from the moment they left the stable Klaas would stand bolt upright, uttering now and then a deep-mouthed bark that became fierce and defiant as the cart approached the gardenhuys. In the arena he seemed to act on Prince Eugene's principle, —that "there is profit in offensive operations." He never waited for an attack, and, being himself a consummate master of that art, never permitted his adversary to take an unfair advantage. Generosity and fear were equally foreign to his nature. A stumbling foe was promptly overthrown, a prostrate one at once torn into pieces. He knew no mercy. He was a perfect beast of prey, and nothing else. And, with all his audacity, he had not the foolhardiness of a bull-dog. Unknown animals he studied before he attacked them. They once pitted him against a wanderoo, or Cingalese baboon, a brute with the face of a gargoyle and the mane of a lion. Klaas kept as still as a mouse, and, with his tail stiffly erect, walked round and round the ring and scrutinized the phenomenon. He did not like the cunning eyes of the half-man, but somehow or other he made up his mind that, whatever the creature might be, he was no fighter, and, slowly contracting his circle, he suddenly and without the least warning sprang upon the stranger and massacred him on the spot, though the monkey nearly pulled the ears out of his head.

Klaas fought about once a month. He did not always come off unscathed, but, though he got sundry desperate rips, he somehow contrived to preserve his anatomic integrity. Three years ago he was as popular as ever, and, unless that Groningen hond has managed to collar him from the left, he probably still holds his levees on the Prinzengraacht.

THE END.

www.ingramcontent.com/pod-product-compliance
Lightning Source LLC
Chambersburg PA
CBHW032145230426
43672CB00011B/2457